OXFORD LOGIC GUIDES

General Editors

ANGUS MACINTYRE
JOHN SHEPHERDSON
DANA SCOTT

OXFORD LOGIC GUIDES

1. Jane Bridge: *Beginning model theory: the completeness theorem and some consequences*
2. Michael Dummett: *Elements of intuitionism*
3. A. S. Troelstra: *Choice sequences: a chapter of intuitionistic mathematics*
4. J. L. Bell: *Boolean-valued models and independence proofs in set theory* (1st edition)
5. Krister Segerberg: *Classical propositional operators: an exercise in the foundation of logic*
6. G. C. Smith: *The Boole–De Morgan correspondence 1842–1864*
7. Alec Fisher: *Formal number theory and computability: a work book*
8. Anand Pillay: *An introduction to stability theory*
9. H. E. Rose: *Subrecursion: functions and hierarchies*
10. Michael Hallett: *Cantorian set theory and limitation of size*
11. R. Mansfield and G. Weitkamp: *Recursive aspects of descriptive set theory*
12. J. L. Bell: *Boolean-valued models and independence proofs in set theory* (2nd edition)
13. Melvin Fitting: *Computability theory: semantics and logic programming*
14. J. L. Bell: *Toposes and local set theories: an introduction*
15. Richard Kaye: *Models of Peano arithmetic*
16. Jonathan Chapman and Frederick Rowbottom: *Relative category theory and geometric morphisms: a logical approach*
17. S. Shapiro: *Foundations without foundationalism*
18. J. P. Cleave: *A study of logics*
19. Raymond M. Smullyan: *Gödel's incompleteness theorems*
20. T. E. Forster: *Set theory with a universal set*
21. C. McLarty: *Elementary categories, elementary toposes*
22. Raymond M. Smullyan: *Recursion theory for metamathematics*

Gödel's Incompleteness Theorems

RAYMOND M. SMULLYAN

New York   Oxford
Oxford University Press
1992

Oxford University Press

Oxford   New York   Toronto
Delhi   Bombay   Calcutta   Madras   Karachi
Kuala Lumpur   Singapore   Hong Kong   Tokyo
Nairobi   Dar es Salaam   Cape Town
Melbourne   Auckland

and associated companies in
Berlin   Ibadan

Copyright © 1992 by Oxford University Press, Inc.

Published by Oxford University Press, Inc.,
200 Madison Avenue, New York, New York 10016

Oxford is a registered trademark of Oxford University Press

All rights reserved. No part of this publication may be reproduced,
stored in a retrieval system, or transmitted in any form or by any means,
electronic, mechanical, photocopying, recording, or otherwise,
without the prior permission of Oxford University Press.

Library of Congress Cataloging-in-Publication Data
Smullyan, Raymond M.
Gödel's incompleteness theorems / Raymond M. Smullyan.
p. cm. (Oxford logic guides : 19)
Includes bibliographical references and index.
ISBN 0-19-504672-2
1. Gödel's theorem.   I. Title.   II. Series.
QA9.65.S69   1992   511.3—dc20
92-16377

9 8 7 6 5 4 3 2 1

Printed in the United States of America
on acid-free paper

To Blanche

Preface

This introduction to Gödel's incompleteness theorems is written for the general mathematician, philosopher, computer scientist and any other curious reader who has at least a nodding acquaintance with the symbolism of first-order logic (the logical connectives and quantifiers) and who can recognize the logical validity of a few elementary formulas. A standard one-semester course in mathematical logic is more than enough for the understanding of this volume.

The proofs we give are unusually simple. The simplest proof of all (and we are now addressing the expert rather than the beginning reader) is obviously the one using Tarski's truth set and is the one we give first. It is so very simple compared to the standard ones that we are surprised that it is not generally better known. [Something of the sort can be found in Mostowski [1952], though the last chapter of Quine [1940] comes closer to what we have in mind.]

In our opening chapter we begin with a simple illustration of Gödel's essential idea (using a simple machine language) and then go on to consider some purely abstract incompleteness theorems suggested by the introduction to Gödel's original paper. We show how any mathematical system having certain very general features is subject to Gödel's argument. In subsequent chapters we consider some specific mathematical systems and show that they *do* possess these general features.

In Chapter 2 we prove Tarski's theorem for arithmetic based on plus, times and power. In the next chapter we give our first proof of Gödel's incompleteness theorem for *axiomatic* arithmetic based on plus, times and power and in Chapter 4 we prove the incompleteness of the better known system of Peano Arithmetic based on plus and times alone. Both proofs use Tarski's truth set, which, as we have remarked, accounts largely for their simplicity. [Briefly, provability is arithmetic; truth is not, hence the two do not coincide.] Another simplifying factor is that we use the Montague-Kalish axiomatization of first-order logic, thus circumventing the necessity of arithmetizing substitution. [In a later chapter we indicate in a series of exercises how the proofs can be modified for the more standard type of axiomatization.]

We then turn to the better known incompleteness proofs which do not use the notion of *truth*—Gödel's original proof based on ω-consistency (Ch. 5) and Rosser's proof based on simple consistency (Ch. 6). In establishing these results, we do not employ the usual apparatus of recursive functions; instead, we use the constructive arithmetic relations introduced in T.F.S. (Theory of Formal Systems). It is particularly easy to show these relations to be definable in the formal theories under consideration and so our proofs of the Gödel theorem and the Gödel-Rosser theorem are accordingly simplified. Another simplifying factor is that instead of working with the characteristic functions of the key metamathematical relations (which is the usual procedure) we work with the relations themselves and show directly their representability in the theories under discussion.

By the end of Chapter 6, the reader will have seen three different proofs of the incompleteness of Peano Arithmetic. Each is of interest and reveals certain facts not revealed by either of the others. The three proofs generalize in different directions which we carefully point out and compare. For sheer directness and simplicity, the Gödel-Tarski proof is the best. For applications to Gödel's second theorem, Gödel's original proof is the one that is needed. Out of Rosser's proof has come Kleene's symmetric form of Gödel's theorem and the whole subject of recursive and effective inseparability—a topic we study in great detail in our sequel to this volume.

Chapter 7 is devoted to the remarkable representation and separation theorems of John Shepherdson [1961]. These results are not necessary for the remaining chapters of this volume, but they are extremely fascinating in their own right (and play a major rôle in our sequel). There is a good deal in this chapter that should interest the expert as well as the general reader (for example, a strengthening of Shepherdson's theorem that has somewhat the flavor of recursion theorems, and also some curious variants of Rosser's undecidable sentence).

Chapter 8 contains some basic technical material and a proof of a fixed point principle necessary for the study of Gödel's second incompleteness theorem and Löb's theorem, which we discuss in Chapter 9. Chapter 10 contains some general observations about provability and truth and the statement of an interesting result of Askanas [1975]. Our closing chapter (a bit of dessert) combines some of the author's typical logic puzzles with a review and generalization of several results of earlier chapters and shows how they tie up with recent developments in modal logic of the sort so skillfully treated in Boolos [1979] (and in a semi-formal way, in Smullyan [1987]).

Although this volume was written primarily as an introduction to incompleteness theorems, it was also intended as a preparation for our sequel, "Recursion Theory for Metamathematics", in which we explore in depth the fascinating interrelations between incompleteness and recursive unsolvability.

I wish to express my thanks to Dana Scott, Anil Gupta, Perry Smith and my students Peter Harlan, Suresh Srivinas and Venkatesh Choppella for having made many extremely helpful suggestions.

Contents

I   The General Idea Behind Gödel's Proof   1
   I.   Abstract Forms of Gödel's and Tarski's Theorems . . .   5
   II. Undecidable Sentences of $\mathcal{L}$ 10

II   Tarski's Theorem for Arithmetic   14
   I.   The Language $\mathcal{L}_E$ 14
      §1.   Syntactic Preliminaries. 14
      §2.   The Notion of Truth in $\mathcal{L}_E$. 17
      §3.   Arithmetic and arithmetic Sets and Relations. 19
   II. Concatenation and Gödel Numbering 20
      §4.   Concatenation to the Base b. 20
      §5.   Gödel Numbering. 22
   III. Tarski's Theorem 24
      §6.   Diagonalization and Gödel Sentences. 24

III   The Incompleteness of Peano Arithmetic With Exponentiation   28
   I.   The Axiom System P.E. 28
      §1.   The Axiom System P.E. 28
   II. Arithmetization of the Axiom System 30
      §2.   Preliminaries. 30
      §3.   Arithmetization of the Syntax of P.E. 33
      §4.   Gödel's Incompleteness Theorem for P. E.. . . 36

IV   Arithmetic Without the Exponential   40
   I.   The Incompleteness of P.A. 40
      §3.   Concatenation to a Prime Base. 43
      §4.   The Finite Set Lemma. 45
      §5.   Proof of Theorem E. 46
      §6.   The Incompleteness of Peano Arithmetic. . . . 49
   II. More on Σ_1-Relations 50

V	**Gödel's Proof Based on ω-Consistency**	**56**
I.	Some Abstract Incompleteness Theorems	58
	§1. A Basic Incompleteness Theorem.	58
	§2. The ω-Consistency Lemma.	61
II.	Σ_0-Completeness	66
	§4. Some Σ_0-complete Subsystems of Peano Arithmetic.	68
	§6. The ω-Incompleteness of P.A.	73
VI	**Rosser Systems**	**75**
	§1. Some Abstract Incompleteness Theorems After Rosser.	76
	§2. A General Separation Principle.	77
	§3. Rosser's Undecidable Sentence.	81
	§4. The Gödel and Rosser Sentences Compared.	82
	§5. More on Separation.	84
VII	**Shepherdson's Representation Theorems**	**86**
	§1. Shepherdson's Representation Theorem.	86
	§2. Exact Rosser Systems.	90
	§3. Some Variants of Rosser's Undecidable Sentence.	93
	§4. A Strengthening of Shepherdson's Theorems	96
VIII	**Definablity and Diagonalization**	**97**
	§1. Definability and Complete Representability.	97
	§2. Strong Definability of Functions in $\mathcal{S}$.	98
	§3. Strong Definablity of Recursive Functions in (R).	100
	§4. Fixed Points and Gödel Sentences.	102
	§5. Truth Predicates.	104
IX	**The Unprovability of Consistency**	**106**
	§1. Provability Predicates.	106
	§2. The Unprovability of Consistency.	108
	§3. Henkin Sentences and Löb's Theorem.	109
X	**Some General Remarks on Provability and Truth**	**112**
XI	**Self-Referential Systems**	**116**
I.	Logicians Who Reason About Themselves	116
	§1. An Analogue of the Tarski-Gödel Theorem.	116
	§2. Normal and Stable Reasoners of Type 1.	117
	§3. Rosser Type Reasoners.	120

	§4.	The Consistency Problem. 122
	§5.	Self-Fulfilling Beliefs and Löb's Theorem. . . . 124
II.	Incompleteness Arguments in a General Setting 126
III.	Systems of Type G 129
IV.	Modal Systems 132

References 136

Index 138

Gödel's Incompleteness Theorems

Chapter I

The General Idea Behind Gödel's Proof

In the next several chapters we will be studying incompleteness proofs for various axiomatizations of arithmetic. Gödel, 1931, carried out his original proof for axiomatic set theory, but the method is equally applicable to axiomatic number theory. The incompleteness of axiomatic number theory is actually a stronger result since it easily yields the incompleteness of axiomatic set theory.

Gödel begins his memorable paper with the following startling words.

"The development of mathematics in the direction of greater precision has led to large areas of it being formalized, so that proofs can be carried out according to a few mechanical rules. The most comprehensive formal systems to date are, on the one hand, the Principia Mathematica of Whitehead and Russell and, on the other hand, the Zermelo-Fraenkel system of axiomatic set theory. Both systems are so extensive that all methods of proof used in mathematics today can be formalized in them—i.e. can be reduced to a few axioms and rules of inference. It would seem reasonable, therefore, to surmise that these axioms and rules of inference are sufficient to decide *all* mathematical questions which can be formulated in the system concerned. In what follows it will be shown that this is not the case, but rather that, in both of the cited systems, there exist relatively simple problems of the theory of ordinary whole numbers which cannot be decided on the basis of the axioms."

Gödel then goes on to explain that the situation does not depend on the special nature of the two systems under consideration but holds for an extensive class of mathematical systems.

Just what is this "extensive class" of mathematical systems? Various interpretations of this phrase have been given, and Gödel's theorem has accordingly been generalized in several ways. We will consider many such generalizations in the course of this volume. Curiously enough, one of the generalizations that is most direct and most easily accessible to the general reader is also the one that appears to be the least well known. What makes this particularly curious is that the way in question is the very one indicated by Gödel himself in the introductory section of his original paper! We shall shortly turn to this (or rather to a further generalization of it), but before that, we would like the reader to look at the following little puzzles which illustrate Gödel's essential idea in a simple and instructive way.

A Gödelian Puzzle. Let us consider a computing machine that prints out various expressions composed of the following five symbols:

$$\sim P\ N\ (\ )$$

By an *expression*, we mean any finite non-empty string of these five symbols. An expression X is called *printable* if the machine can print it. We assume the machine programmed so that any expression that the machine can print will be printed sooner or later.

By the *norm* of an expression X, we shall mean the expression $X(X)$—e.g. the norm of $P\sim$ is $P\sim(P\sim)$. By a *sentence*, we mean any expression of one of the following four forms (X is any expression):

(1) $P(X)$
(2) $PN(X)$
(3) $\sim P(X)$
(4) $\sim PN(X)$

Informally, P stands for "printable"; N stands for "the norm of" and $\sim$ stands for "not". And so we define $P(X)$ to be *true* if (and only if) X is printable. We define $PN(X)$ to be true if the *norm* of X is printable. We call $\sim P(X)$ true iff (if and only if) X is not printable, and $\sim PN(X)$ is defined to be true iff the norm of X is not printable. [This last sentence we read as "Not printable the norm of X", or, in better English: "The norm of X is not printable".]

We have now given a perfectly precise definition of what it means for a sentence to be true, and we have here an interesting case of self-reference: The machine is printing out various sentences about what the machine can and cannot print, and so it is describing its own behavior! [It somewhat resembles a self-conscious organism, and we can see why such computers are of interest to those working in artificial intelligence.]

We are given that the machine is completely accurate in that all sentences printed by the machine are true. And so, for example, if the machine ever prints $P(X)$, then X really is printable (X will be printed by the machine sooner or later). Also, if $PN(X)$ is printable, so is $X(X)$ (the norm of X). Now, suppose X is printable. Does it follow that $P(X)$ is printable? Not necessarily. If X is printable, then $P(X)$ is certainly *true*, but we are not given that the machine is capable of printing *all* true sentences but only that the machine never prints any false ones. [Whether the machine can print expressions that are not sentences at all is immaterial. The important thing is that among the *sentences* printable by the machine, all of them are true.]

Is it possible that the machine *can* print all true sentences? The answer is *no* and the problem for the reader is this: Find a true sentence that the machine cannot print. [Hint: Find a sentence that asserts its own non-printability—i.e. one which is true if and only if it is not printable by the machine. The solution is given after the next problem.]

A Variant of the Puzzle. The following variant of the above puzzle will introduce the reader to the notion of *Gödel numbering*.

We now have another machine that prints out expressions composed of the following five symbols:

$$\sim \quad P \quad N \quad 1 \quad 0$$

We are representing the natural numbers in binary notation (as strings of 1's and 0's), and for purposes of this problem, we will identify the natural numbers with the binary numerals that represent them.

To each expression we assign a number which we call the *Gödel number* of the expression. We do this according to the following scheme: The individual symbols $\sim, P, N, 1, 0$ are assigned the respective Gödel numbers $10, 100, 1000, 10000, 100000$. Then, the Gödel number of a compound expression is obtained by replacing each symbol by *its* Gödel number—for example, PNP has Gödel number 1001000100. We redefine the *norm* of an expression to be the expression followed by its Gödel number—for example, the norm of PNP is the expression $PNP1001000100$. A *sentence* is now an expression of one of the four forms: $PX, PNX, \sim PX$ and $\sim PNX$, where X is any number (written in binary notation). We call PX true if X is the *Gödel number* of a printable expression. We call PNX true iff X is the Gödel number of an expression whose *norm* is printable. We call $\sim PX$ true if PX is not true (X is not the Gödel number of a printable expression), and we call $\sim PNX$ true iff PNX is not true.

Again we are given that the machine never prints a false sentence. Find a true sentence that the machine cannot print.

Solutions. For the first problem, the sentence is $\sim PN(\sim PN)$. By definition of "true", this sentence is true if and only if the norm of $\sim PN$ is not printable. But the norm of $\sim PN$ is the very sentence $\sim PN(\sim PN)$! And so the sentence is true if and only if it is not printable. This means that either the sentence is true and not printable, or it is printable and not true. The latter alternative violates the given hypothesis that the machine never prints sentences that are not true. Hence the sentence must be true, but the machine cannot print it.

Of course, instead of having talked about a machine that *prints* various expressions in our five symbols, we could have talked about a mathematical system that *proves* various sentences in the same five symbols. We would then reinterpret the letter P to mean *provable* in the system, rather than printable by the machine. Then, given that the system is wholly accurate (in that, false sentences are never provable in it), the sentence $\sim PN(\sim PN)$ would be a sentence that is true but not provable in the system.

Let us further observe that the sentence $PN(\sim PN)$ is false (since its negation is true). Hence it is also not provable in the system (assuming that the system is accurate). And so the sentence

$$PN(\sim PN)$$

is an example of a sentence *undecidable* in a system—i.e. neither it nor its negation is provable in the system.

For the second problem, the solution is $\sim PN101001000$.

Now we shall turn to some incompleteness arguments in a general setting: We consider a very broad notion of a mathematical system and show that if it has certain features, then Gödel's argument goes through. In the chapters that follow, we will look at some particular systems and show that they do indeed possess these features.

I. Abstract Forms of Gödel's and Tarski's Theorems

Each of the languages $\mathcal{L}$ to which Gödel's argument is applicable contains at least the following items.

1. A denumerable set $\mathcal{E}$ whose elements are called the *expressions* of $\mathcal{L}$.
2. A subset $\mathcal{S}$ of $\mathcal{E}$ whose elements are called the *sentences* of $\mathcal{L}$.
3. A subset $\mathcal{P}$ of $\mathcal{S}$ whose elements are called the *provable* sentences of $\mathcal{L}$.
4. A subset $\mathcal{R}$ of $\mathcal{S}$ whose elements are called the *refutable* (sometimes *disprovable*) sentences of $\mathcal{L}$.
5. A set $\mathcal{H}$ of expressions whose elements are called the *predicates* of $\mathcal{L}$. [These were called *class names* in Gödel's introduction. Informally, each predicate H is thought of as being the name of a set of natural numbers.]
6. A function Φ that assigns to every expression E and every natural number n an expression $E(n)$. The function is required to obey the condition that for every predicate H and every natural number n, the expression $H(n)$ is a sentence. [Informally, the sentence $H(n)$ expresses the proposition that the number n belongs to the set named by H.]

In the first incompleteness proof that we will give for a *particular* system $\mathcal{L}$, we will use a basic concept made precise by Alfred Tarski [1936]—viz. the notion of a *true* sentence (defined quite differently than that of a provable sentence of a system). And so we consider a seventh and final item of our language $\mathcal{L}$.

7. A set $\mathcal{T}$ of sentences whose elements are called the *true* sentences of $\mathcal{L}$.

This concludes our abstract description of the type of systems that we will study in the next several chapters.

Expressibility in $\mathcal{L}$. The notion of *expressibility* in $\mathcal{L}$, which we are about to define, concerns the truth set $\mathcal{T}$ but does not concern either of the sets $\mathcal{P}$ and $\mathcal{R}$.

The word *number* shall mean *natural number* for the rest of this volume. We will say that a predicate H is *true* for a number n or that n *satisfies* H if $H(n)$ is a true sentence (i.e. is an element of $\mathcal{T}$). By the set *expressed* by H, we mean the set of all n that satisfy

H. Thus for any set A of numbers, H expresses A if and only if for every number n:

$$H(n) \in \mathcal{T} \leftrightarrow n \in A.$$

Definition. A set A is called *expressible* or *nameable* in $\mathcal{L}$ if A is expressed by some predicate of $\mathcal{L}$.

Since there are only denumerably many expressions of $\mathcal{L}$, then there are only finitely or denumerably many predicates of $\mathcal{L}$. But by Cantor's well-known theorem, there are non-denumerably many sets of natural numbers. Therefore, not every set of numbers is expressible in $\mathcal{L}$.

Definition. The system $\mathcal{L}$ is called *correct* if every provable sentence is true and every refutable sentence is false (not true). This means that $\mathcal{P}$ is a subset of $\mathcal{T}$ and $\mathcal{R}$ is disjoint from $\mathcal{T}$. We are now interested in sufficient conditions that $\mathcal{L}$, if correct, must contain a true sentence not provable in $\mathcal{L}$.

Gödel Numbering and Diagonalization. We let g be a 1-1 function which assigns to each expression E a natural number $g(E)$ called the *Gödel number* of E. The function g will be constant for the rest of this chapter. [In the concrete systems to be studied in subsequent chapters, a specific Gödel numbering will be given. Our present purely abstract treatment, however, applies to an arbitrary Gödel numbering.] It will be technically convenient to assume that every number is the Gödel number of an expression. [Gödel's original numbering did not have this property, but the Gödel numbering we will use in subsequent chapters will have this property. However, the results of this chapter can, with minor modifications, be proved without this restriction (cf. Ex. 5).] Assuming now that every number n is the Gödel number of a unique expression, we let E_n be that expression whose Gödel number is n. Thus, $g(E_n) = n$.

By the *diagonalization* of E_n we will mean the expression $E_n(n)$. If E_n is a predicate, then its diagonalization is, of course, a sentence; this sentence is true iff the predicate E_n is satisfied by its own Gödel number n. [We write "iff" to mean if and only if; we use "$\leftrightarrow$" synonymously.]

For any n, we let $d(n)$ be the Gödel number of $E_n(n)$. The function $d(x)$ plays a key rôle in all that follows; we call it the *diagonal function* of the system.

We use the term *number-set* to mean set of (natural) numbers. For any number set A, by A^* we shall mean the set of all numbers

I. Abstract Forms of Gödel's and Tarski's Theorems

n such that $d(n) \in A$. Thus for any n, the equivalence

$$n \in A^* \leftrightarrow d(n) \in A$$

holds by definition of A^*. [A^* could also be written $d^{-1}(A)$, since it is the inverse image of A under the diagonal function $d(x)$.]

An Abstract Form of Gödel's Theorem. We let P be the set of Gödel numbers of all the provable sentences. For any number set A, by its complement $\tilde{A}$, we mean the complement of A relative to the set N of natural numbers—i.e. $\tilde{A}$ is the set of all natural numbers not in A.

Theorem (GT)—After Gödel with shades of Tarski. *If the set $\tilde{P}^*$ is expressible in $\mathcal{L}$ and $\mathcal{L}$ is correct, then there is a true sentence of $\mathcal{L}$ not provable in $\mathcal{L}$.*

Proof. Suppose $\mathcal{L}$ is correct and $\tilde{P}^*$ is expressible in $\mathcal{L}$. Let H be a predicate that expresses $\tilde{P}^*$ in $\mathcal{L}$, and let h be the Gödel number of H. Let G be the diagonalization of H (i.e. the sentence $H(h)$). We will show that G is true but not provable in $\mathcal{L}$.

Since H expresses $\tilde{P}^*$ in $\mathcal{L}$, then for any number n, $H(n)$ is *true* $\leftrightarrow$ $n \in \tilde{P}^*$. Since this equivalence holds for *every* n, then it holds in particular for n the number h. So we take h for n (and this is the part of the argument called *diagonalizing*) and we have the equivalence: $H(h)$ is *true* $\leftrightarrow h \in \tilde{P}^*$. Now,

$$h \in \tilde{P}^* \leftrightarrow d(h) \in \tilde{P} \leftrightarrow d(h) \notin P.$$

But $d(h)$ is the Gödel number of $H(h)$ (since h is the Gödel number of H) and so $d(h) \in P \leftrightarrow H(h)$ is provable in $\mathcal{L}$ and $d(h) \notin P \leftrightarrow H(h)$ is not provable in $\mathcal{L}$. And so we have

1. $H(h)$ is *true* $\leftrightarrow H(h)$ is not provable in $\mathcal{L}$. This means that $H(h)$ is either true and not provable in $\mathcal{L}$ or false but provable in $\mathcal{L}$. The latter alternative violates the hypothesis that $\mathcal{L}$ is correct. Hence it must be that $H(h)$ is true but not provable in $\mathcal{L}$.

When it comes to the particular languages $\mathcal{L}$ that we will study, we will verify the hypothesis that $\tilde{P}^*$ is expressible in $\mathcal{L}$ by separately verifying the following three conditions.

G_1: For any set A expressible in $\mathcal{L}$, the set A^* is expressible in $\mathcal{L}$.
G_2: For any set A expressible in $\mathcal{L}$, the set $\tilde{A}$ is expressible in $\mathcal{L}$.
G_3: The set P is expressible in $\mathcal{L}$.

Conditions G_1 and G_2, of course, imply that for any set A expressible in $\mathcal{L}$, the set $\tilde{A}^*$ is expressible in $\mathcal{L}$. Hence if P is expressible in $\mathcal{L}$, then so is $\tilde{P}^*$.

We might remark that the verification of G_1 will turn out to be relatively simple; the verification of G_2 will be completely trivial; but the verification of G_3 will turn out to be extremely elaborate.

Gödel Sentences. Woven into the proof of Theorem GT is a very important principle which was made explicit by Rudolf Carnap [1934] and which is closely related to Tarski's theorem, to which we will soon turn.

Call a sentence E_n a *Gödel* sentence for a number set A if either E_n is true and its Gödel number n lies in A, or E_n is false and its Gödel number lies outside A. Thus, E_n is a Gödel sentence for A iff the following condition holds:

$$E_n \in \mathcal{T} \leftrightarrow n \in A.$$

[Informally, a Gödel sentence for A can be thought of as a sentence asserting that its own Gödel number lies in A. If the sentence is true, then its Gödel number does lie in A. If the sentence if false, then its Gödel number does not lie in A.]

The following lemma and theorem pertains only to the set $\mathcal{T}$. The sets $\mathcal{P}$ and $\mathcal{R}$ are irrelevant.

Lemma (D)—A Diagonal Lemma. (a) For any set A, if A^* is expressible in $\mathcal{L}$, then there is a Gödel sentence for A.

(b) If $\mathcal{L}$ satisfies condition G_1, then for any set A expressible in $\mathcal{L}$, there is a Gödel sentence for A.

Proof.

(a) Suppose H is a predicate that expresses A^* in $\mathcal{L}$; let h be its Gödel number. Then $d(h)$ is the Gödel number of $H(h)$. For any number n, $H(n)$ is *true* $\leftrightarrow n \in A^*$, therefore, $H(h)$ is *true* $\leftrightarrow h \in A^*$. And $h \in A^* \leftrightarrow d(h) \in A$. Therefore, $H(h)$ is *true* $\leftrightarrow d(h) \in A$, and since $d(h)$ is the Gödel number of $H(h)$, then $H(h)$ is a Gödel sentence for A.

(b) Immediate from (a).

Let us note that if we had first proved Lemma D, we would have had the following swift proof of Theorem GT: Since $\tilde{P}^*$ is nameable in $\mathcal{L}$, then by lemma D, there is a Gödel sentence G for $\tilde{P}$. A Gödel sentence for $\tilde{P}$ is nothing more nor less than a sentence which is true if and only if it is not provable (in $\mathcal{L}$). And for any *correct* system $\mathcal{L}$, a Gödel sentence for $\tilde{P}$ is a sentence which is true but not provable in $\mathcal{L}$. [Such a sentence can be thought of as asserting its own non-provability in $\mathcal{L}$.]

An Abstract Form of Tarski's Theorem. Lemma D has another important consequence: Let T be the set of Gödel numbers of the *true* sentences of $\mathcal{L}$. Then the following theorem holds.

Theorem (T) (After Tarski).

1. *The set $\tilde{T}^*$ is not nameable in $\mathcal{L}$.*
2. *If condition G_1 holds, then $\tilde{T}$ is not nameable in $\mathcal{L}$.*
3. *If conditions G_1 and G_2 both hold, then the set T is not nameable in $\mathcal{L}$.*

Proof. To begin with, there cannot possibly be a Gödel sentence for the set $\tilde{T}$ because such a sentence would be true if and only if its Gödel number was *not* the Gödel number of a true sentence, and this is absurd.

1. If $\tilde{T}^*$ were nameable in $\mathcal{L}$, then by (a) of Lemma D, there would be a Gödel sentence for the set $\tilde{T}$, which we have just shown is impossible. Therefore, $\tilde{T}^*$ is not nameable in $\mathcal{L}$.
2. Suppose condition G_1 holds. Then if $\tilde{T}$ were nameable in $\mathcal{L}$, the set $\tilde{T}^*$ would be nameable in $\mathcal{L}$, violating (1).
3. If G_2 also holds, then if T were nameable in $\mathcal{L}$, then $\tilde{T}$ would also be nameable in $\mathcal{L}$, violating (2).

Remarks.

1. Conclusion (3) above is sometimes paraphrased: For systems of sufficient strength, truth within the system is not definable within the system. The phrase "sufficient strength" has been interpreted in several ways. We would like to point out that conditions G_1 and G_2 suffice for this "sufficient strength."
2. Gödel (1931) likens his proof to the famous paradox of the Cretan who says that all Cretans are liars.[1] An analogy that comes closer to Gödel's theorem is this: Imagine a land in which every inhabitant either always tells the truth or always lies. Some of the inhabitants are Athenians and some are Cretans. It is given

---

[1] Actually, the liar paradox is more closely related to Tarski's theorem than to Gödel's.

that all the Athenians of the land always tell the truth and all the Cretans of the land always lie. What statement could an inhabitant make that would convince you that he always tells the truth but that he is not an Athenian?

All he would need to say is: "I am not an Athenian." A liar couldn't make that claim (because a liar is really *not* an Athenian; only truth-tellers are Athenian). Therefore, he must be truthful. Hence his statement was true, which means that he is really not an Athenian. So he is a truth teller but not an Athenian.

If we think of the Athenians as playing the rôle of the sentences of $\mathcal{L}$, which are not only true but provable in $\mathcal{L}$, then any inhabitant who claims he is not Athenian plays the rôle of Gödel's sentence G, which asserts its own non-provability in $\mathcal{L}$. [The Cretans, of course, play the rôle of the refutable sentences of $\mathcal{L}$, but their function won't emerge till a bit later.]

II. Undecidable Sentences of $\mathcal{L}$

So far, the set $\mathcal{R}$ of *refutable* sentences has played no rôle. Now it shall play a key one.

$\mathcal{L}$ is called *consistent* if no sentence is both provable and refutable in $\mathcal{L}$ (i.e. the sets $\mathcal{P}$ and $\mathcal{R}$ are disjoint) and *inconsistent* otherwise. The definition of consistency refers only to the sets $\mathcal{P}$ and $\mathcal{R}$, not to the set $\mathcal{T}$. Nevertheless, if $\mathcal{L}$ is correct, then it is automatically consistent (because if $\mathcal{P}$ is a subset of $\mathcal{T}$ and $\mathcal{T}$ is disjoint from $\mathcal{R}$, then $\mathcal{P}$ must be disjoint from $\mathcal{R}$). The converse is not necessarily true (we will later consider some systems that are consistent but not correct).

A sentence X is called *decidable* in $\mathcal{L}$ if it is either provable or refutable in $\mathcal{L}$ and *undecidable* in $\mathcal{L}$ otherwise. The system $\mathcal{L}$ is called *complete* if every sentence is decidable in $\mathcal{L}$ and *incomplete* if some sentence is undecidable in $\mathcal{L}$.

Suppose now $\mathcal{L}$ satisfies the hypothesis of Theorem GT. Then some sentence G is true but not provable in $\mathcal{L}$. Since G is true, it is not refutable in $\mathcal{L}$ either (by the assumption of correctness). Hence G is undecidable in $\mathcal{L}$. And so we at once have

II. Undecidable Sentences of $\mathcal{L}$

Theorem 1. If $\mathcal{L}$ is correct and if the set $\tilde{P}^*$ is expressible in $\mathcal{L}$, then $\mathcal{L}$ is incomplete.

A Dual of Theorem 1. In T.F.S. (Theory of Formal Systems, 1961) we introduced what might aptly be called a "dual form" of Gödel's argument, which we will first explain informally. Instead of constructing a sentence that says "I am not provable," we will construct a sentence that says "I *am* refutable." As we are about to see, such a sentence must also be undecidable in $\mathcal{L}$ (if $\mathcal{L}$ is correct).

We have defined P to be the set of Gödel numbers of the provable sentences. We now define R to be the set of Gödel numbers of the refutable sentences.

Theorem (1°)—(A Dual of Theorem 1). If $\mathcal{L}$ is correct and the set R^* is expressible in $\mathcal{L}$, then $\mathcal{L}$ is incomplete. More specifically, if $\mathcal{L}$ is correct and K is a predicate that expresses the set R^*, then its diagonalization $K(k)$ is undecidable in $\mathcal{L}$ (k is the Gödel number of K).

Proof. Assume hypothesis. Since K expresses R^*, then by the proof of (a) of Lemma D, the sentence $K(k)$ is a Gödel sentence for the set R. Thus, $K(k)$ is true iff its Gödel number is in R, or, what is the same thing, $K(k)$ is true iff $K(k)$ is in $\mathcal{R}$, so $K(k)$ is true iff $K(k)$ is refutable in $\mathcal{L}$. This means that $K(k)$ is either true and refutable or false but not refutable. By the assumption of correctness, $K(k)$ cannot be true and refutable. Hence it is false but not refutable. Since the sentence is false, it is not provable either (again by the assumption that $\mathcal{L}$ is correct). Hence $K(k)$ is neither provable nor refutable in $\mathcal{L}$.

Remarks. Just as the Gödel sentence $H(h)$ can be thought of as saying: "I am not provable in $\mathcal{L}$," we can think of $K(k)$ as saying: "I am refutable in $\mathcal{L}$." Going back to our analogy of Athenians and Cretans, just as $H(h)$ corresponds to an inhabitant who claims that he is not an Athenian, the sentence $K(k)$ corresponds to an inhabitant who claims that he *is* a Cretan. He must be a liar but not a Cretan. Hence (like an inhabitant who claims he is not an Athenian) he must be neither an Athenian nor a Cretan.

Suppose now we have a correct system $\mathcal{L}$ satisfying the following two conditions:
G_1: For any expressible set A, the set A^* is expressible
G_3': The set R is expressible
Then, of course, the set R^* is expressible. So by Theorem 1°, $\mathcal{L}$

is inconsistent or incomplete. We note that the complementation condition G_2 is not required in this proof.

The first exercise below contains an interesting variant of Theorem 1°.

Exercise 1. Suppose $\mathcal{L}$ is a correct system such that the following two conditions hold.

1. The set P^* is expressible in $\mathcal{L}$.
2. For any predicate H, there is a predicate H' such that for every n, the sentence $H'(n)$ is provable in $\mathcal{L}$ if and only if $H(n)$ is refutable in $\mathcal{L}$.

Prove that $\mathcal{L}$ is incomplete.

Exercise 2. We say that a predicate H *represents* a set A in $\mathcal{L}$ if for every number n, the sentence $H(n)$ is *provable* in $\mathcal{L}$ if and only if $n \in A$. [Note that this definition makes no reference to the truth set $\mathcal{T}$ but only to the provability set $\mathcal{P}$.]

Show that if the set R^* is representable in $\mathcal{L}$, then $\mathcal{L}$, if consistent, is incomplete.

Exercise 3. Show that if some superset of R^* disjoint from P^* is representable in $\mathcal{L}$, then $\mathcal{L}$ is incomplete. [We call B a *superset* of A if A is a subset of B.]

Exercise 4. Let us say that a predicate H *contrarepresents* a set A in $\mathcal{L}$ if for every number n, the sentence $H(n)$ is *refutable* in $\mathcal{L}$ iff $n \in A$. Show that if the set P^* is contrarepresentable in $\mathcal{L}$ and $\mathcal{L}$ is consistent, then $\mathcal{L}$ is incomplete. [This result and the result of Exercise 2 will be expanded in Chapter 5; the result of Exercise 3 is related to Rosser's incompleteness proof, which we will study in Chapter 6.]

Exercise 5. Suppose we have a Gödel numbering g such that it is not the case that every number is a Gödel number. Then we define a function $d(x)$ to be *a* diagonal function (rather than *the* diagonal function) if it has the property that for any number e, *if* e is the Gödel number of an expression E, then $d(e)$ is the Gödel number of $E(e)$. Prove that for any diagonal function $d(x)$, if $d^{-1}(A)$ is expressible in $\mathcal{L}$, then there is a Gödel sentence for A.

Exercise 6. Is it necessarily true that for any set A, the set $\widetilde{A}^*$ is the same as the set $\widetilde{A^*}$?

Exercise 7. To emphasize the wholly constructive nature of Gödel's

II. Undecidable Sentences of $\mathcal{L}$

proof, suppose $\mathcal{L}$ is a correct system such that the following three conditions hold.

1. E_7 is a predicate that expresses the set P.
2. For any number n, if E_n is a predicate, then so is E_{3n}, and E_{3n} expresses the complement of the set expressed by E_n.
3. For any number n, if E_n is a predicate, then E_{3n+1} is a predicate, and if A is the set expressed by E_n, then A^* is the set expressed by E_{3n+1}.

   (a) Find numbers a and b (either the same or different) such that $E_a(b)$ is a true sentence not provable in $\mathcal{L}$. [There are two solutions in which a and b are both less than 100. Can the reader find them both?]
   (b) Show that there are infinitely many pairs (a, b) such that $E_a(b)$ is true but not provable in $\mathcal{L}$.
   (c) Given that E_{10} is a predicate, find numbers c and d such that $E_c(d)$ is a Gödel sentence for the set expressed by E_{10}.

Chapter II

Tarski's Theorem for Arithmetic

In the last chapter, we dealt with mathematical languages in considerable generality. We shall now turn to some particular mathematical languages. One of our goals is to reach Gödel's incompleteness theorem for the particular system known as *Peano Arithmetic*. We shall give several proofs of this important result; the simplest one is based partly on Tarski's theorem, to which we first turn.

I. The Language $\mathcal{L}_E$

§1. Syntactic Preliminaries. The first concrete language that we will study is the language of first order arithmetic based on addition, multiplication and exponentiation. [We also take as primitive the successor function and the less than or equal to relation, but these are inessential.] We shall formulate the language using only a finite alphabet (mainly for purposes of a convenient Gödel numbering); specifically we use the following 13 symbols.

$$0 \; ' \; (\;) \; f \; , \; v \; \sim \; \supset \; \forall \; = \; \leq \; \#$$

The expressions $0, 0', 0'', 0''', \cdots$ are called *numerals* and will serve as formal names of the respective natural numbers $0, 1, 2, 3, \cdots$. The accent symbol (also called the *prime*) is serving as a name of the successor function. We also need names for the operations of addition, multiplication and exponentiation; we use the expressions $f_\prime, f_{\prime\prime}, f_{\prime\prime\prime}$ as respective names of these three functions. We abbreviate $f_\prime$ by the familiar "+"; we abbreviate $f_{\prime\prime}$ by the familiar dot and $f_{\prime\prime\prime}$ by the symbol "E".

The symbols $\sim$ and $\supset$ are the familiar symbols from propositional logic, standing for negation and material implication, respectively. [For any reader not familiar with the use of the horseshoe symbol, for any propositions p and q, the propositions $p \supset q$ is intended to mean nothing more nor less than that either p is false, or p and q are both true.] The symbol $\forall$ is the *universal quantifier* and means "for all." We will be quantifying only over natural numbers not over sets or relations on the natural numbers. [Technically, we are working in first-order arithmetic, not second-order arithmetic.]

The symbol "=" is used, as usual, to denote the identity relation, and "$\leq$" is used, as usual, to denote the "less than or equal to" relation.

We also need denumerably many expressions $v_1, v_2, \ldots, v_n, \ldots$ called (individual) *variables*. Well, we wish to stay within our 13-symbol alphabet, and so we will take for $v_1, v_2, v_3, \ldots$ the respective expressions $(v\prime), (v\prime\prime), (v\prime\prime\prime), \ldots$ [Thus, v_n is, by definition, the result of enclosing "v" followed by n subscripts in parentheses.]

Terms and Formulas. An expression is called a *term* if its being so is a consequence of the following two rules.

1. Every variable and numeral is a term.
2. If t_1 and t_2 are terms, then so are $(t_1 + t_2), (t_1 \cdot t_2), (t_1 \text{ E } t_2)$ and t_1'.

A term is said to be a *closed* term or a *constant* term if it contains no variables.

By an *atomic formula*, we mean any expression of one of the two forms $t_1 = t_2$ and $t_1 \leq t_2$, where t_1 and t_2 are any terms. The set of *formulas* is inductively defined by the rules:

1. Every atomic formula is a formula.
2. If F and G are formulas, then $\sim F$ and $(F \supset G)$ are formulas, and for every variable v_i, the expression $\forall v_i F$ is a formula.

Free and Bound Occurrences of Variables. Let v_i be any variable. For any term t, all occurrences of v_i in t are called *free occurrences*. Also for any *atomic* formula A, all occurrences of v_i in A are called free occurrences. For any formulas F and G, the free occurrences of v_i in $(F \supset G)$ are those in F together with those in G. The free occurrences of v_i in $\sim F$ are those in F. Now, v_i has no free occurrences in $\forall v_i F$; all occurrences of v_i in $\forall v_i F$ are called *bound* occurrences. For any $j \neq i$, the free occurrences of v_i in $\forall v_j F$ are those of F.

Sentences. By a *sentence*, we mean any formula in which no variable has any free occurrence. Sentences are sometimes called *closed formulas*. By an *open* formula, we mean a formula which is not closed (i.e., at least one variable has at least one free occurrence in it).

Substitution of Numerals for Variables. For any natural number n, by $\bar{n}$ we mean the numeral designating n (i.e., the symbol "0" followed by n accents). [For example, $\bar{5}$ is the expression $0'''''$.]

For any variable v_i, we sometimes write $F(v_i)$ to mean any formula in which v_i is the only free variable, in which case, by $F(\bar{n})$ we mean the result of substituting the numeral $\bar{n}$ for every free occurrence of v_i in $F(v_i)$. More generally, we write $F(v_{i_1}, \ldots, v_{i_n})$ for any formula in which $v_{i_1}, \ldots, v_{i_n}$ are the only free variables, and for any numbers $k_1, \ldots, k_n$, $F(\bar{k}_1, \ldots, \bar{k}_n)$ is understood to be the result of substituting $\bar{k}_1, \ldots, \bar{k}_n$ for all free occurrences of $v_{i_1}, \ldots, v_{i_n}$ respectively. We call the sentence, $F(\bar{k}_1, \ldots, \bar{k}_n)$, an *instance* of the formula $F(v_{i_1}, \ldots, v_{i_n})$.

We call a formula, $F(v_{i_1}, \ldots, v_{i_n})$, *regular* if $i_1 = 1, \ldots, i_n = n$. Thus, a regular formula F is one such that for any i, if v_i is a free variable of F, then for any $j \leq i$, v_j is also a free variable of F. Thus, a regular formula can be written as $F(v_1, \ldots, v_n)$.

Degrees and Induction. By the *degree* of a formula, we mean the number of occurrences of the logical connectives $\sim$ and $\supset$ and the quantifier $\forall$. Thus,

1. Atomic formulas are of degree 0.
2. For any formulas F and G of respective degrees d_1 and d_2, the formula $\sim F$ is of degree $d_1 + 1$; the formula $(F_1 \supset F_2)$ is of degree $d_1 + d_2 + 1$, and for any variable v_i, the formula $\forall v_i F_1$ is of degree $d_1 + 1$.

We presume familiarity with the principle of mathematical induction from which it follows that to show that a given property holds for all formulas, it suffices to show that it holds for all atomic formulas and that for any formula F, if the property holds for all formulas of lower degree than F, then it also holds for F.

Abbreviations. We employ the following standard abbreviations, where F, F_1 and F_2 are formulas, v_i is any variable, and t_1 and t_2

are any terms.

$$
\begin{aligned}
(F_1 \vee F_2) &\equiv_{\text{df}} (\sim F_1 \supset F_2) \\
(F_1 \wedge F_2) &\equiv_{\text{df}} \sim (F_1 \supset \sim F_2) \\
F_1 \equiv F_2 &\equiv_{\text{df}} ((F_1 \supset F_2) \wedge (F_2 \supset F_1)) \\
\exists v_i F &\equiv_{\text{df}} \sim \forall v_i \sim F \\
t_1 \neq t_2 &\equiv_{\text{df}} \sim t_1 = t_2 \\
t_1 < t_2 &\equiv_{\text{df}} ((t_1 \leq t_2) \wedge (\sim t_1 = t_2)) \\
t_1^{t_2} &\equiv_{\text{df}} t_1 \text{ E } t_2 \\
(\forall v_i \leq t)F &\equiv_{\text{df}} \forall v_i (v_i \leq t \supset F) \\
(\exists v_i \leq t)F &\equiv_{\text{df}} \sim (\forall v_i \leq t) \sim F.
\end{aligned}
$$

In displaying formulas and terms, we shall often omit parentheses if no ambiguity can result. For example, in displaying a formula or term standing alone, we can drop the outermost parentheses—e.g. we can write $F \supset G$ instead of $(F \supset G)$; also the term $(v_1 + v_2)$ standing alone can be abbreviated $v_1 + v_2$, and $((v_1 + v_2) \cdot v_3)$ could be abbreviated $(v_1 + v_2) \cdot v_3$.

Designation. We recall that by a *constant* term, we mean a term with no variables. Each constant term c designates a unique natural number in accordance with the following rules.

1. A numeral $\bar{n}$ designates n.
2. If c_1 and c_2 designates n_1 and n_2 respectively, then $(c_1 + c_2)$ designates the sum of n_1 and n_2; $(c_1 \cdot c_2)$ designates the product of n_1 and n_2; $(c_1 \text{ E } c_2)$ designates the number $n_1^{n_2}$; and $c_1{}'$ designates $n_1 + 1$.

As an example, the constant term $((0'''+0') \cdot (0'' \text{ E } 0'''))'$ designates the number $(4 \cdot 2^3) + 1$, or 33.

§2. The Notion of Truth in $\mathcal{L}_E$.

We now wish to define what it means for a sentence of $\mathcal{L}_E$ to be *true*. The definition will be by induction on the degrees of sentences. The following conditions provide an inductive definition of truth.

T_0: 1. An atomic sentence $c_1 = c_2$ (c_1 and c_2 are constant terms) is true iff c_1 and c_2 designate the same natural number (under the designation rules given above).

    2. An atomic sentence $c_1 \leq c_2$ is true iff the number designated by c_1 is less than or equal to the number designated by c_2.

T_1: A sentence of the form $\sim X$ is true iff X is not true.

T_2: A conditional sentence $X \supset Y$ is true iff either X is not true or both X and Y are true.

T_3: A universal sentence $\forall v_i F$ is true iff for every number n, the sentence $F(\overline{n})$ is true.

Condition T_0 states the truth conditions for atomic sentences outright. Conditions T_1, T_2 and T_3 define the truth of a non-atomic sentence in terms of the truths of all sentences of lower degree. [Note that in T_3, F is of lower degree than $\forall v_i F$. Hence for every n, $F(\overline{n})$ is of lower degree than $\forall v_i F$. Since $\forall v_i F$ is a *sentence*, no variable other than v_i is free in F, and so $F(\overline{n})$ is also a sentence.]

An open formula $F(v_{i_1}, \ldots, v_{i_k})$ cannot be said to be true or false, but we call the formula *correct* if for all numbers $n_1, \ldots, n_k$, the sentence $F(\overline{n}_1, \ldots, \overline{n}_k)$ is true.

Exercise 1.

1. Show that for any sentences X and Y, the sentence $X \wedge Y$ is true iff X and Y are both true.
2. Show that $X \vee Y$ is true iff at least one of X and Y is true.
3. For any formula F in which v_i is the only free variable, show that the sentence $\exists v_i F$ is true iff there is at least one number n such that $F(\overline{n})$ is true.

Substitution of Variables. Consider a formula $F(v_1)$ with v_1 as the only free variable. For any variable v_i ($i \neq 1$), we define $F(v_i)$ as follows:

1. If v_i does not occur as a bound variable of F, then $F(v_i)$ is the result of substituting v_i for all free occurrences of v_1 in F.
2. If v_i does occur as a bound variable of F, we take the smallest number j such that v_j does not occur in F, then we substitute v_j for all occurrences of v_i in F—call this formula $F'(v_j)$—and then we substitute v_i for all free occurrences of v_1 in $F'(v_j)$. This formula we call $F(v_i)$.

For example, let $F(v_1)$ be the formula $\exists v_2 (v_2 \neq v_1)$. This formula is correct. By $F(v_2)$ we mean not the obviously incorrect formula $\exists v_2 (v_2 \neq v_2)$, but the correct formula $\exists v_3 (v_3 \neq v_2)$.

A similar definition applies to regular formulas $F(v_1, \ldots, v_n)$ with more than one free variable. For any variables $v_{i_1}, \ldots, v_{i_n}$, we first rewrite all bound variables of F so that none of $v_{i_1}, \ldots, v_{i_n}$ occurs as a bound variable and then we simultaneously substitute $v_{i_1}, \ldots, v_{i_n}$ for the free occurrences of $v_1, \ldots, v_n$, and we denote this formula by $F(v_{i_1}, \ldots, v_{i_n})$.

Two sentences are called (arithmetically) *equivalent* iff they are either both true or both false. Two open formulas, $F(v_{i_1}, \ldots, v_{i_k})$ and $G(v_{i_1} \ldots, v_{i_k})$, with the same free variables are called equivalent iff for all numbers $n_1, \ldots, n_k$, the sentences $F(\overline{n}_1, \ldots, \overline{n}_k)$ and $G(\overline{n}_1, \ldots, \overline{n}_k)$ are equivalent.

§3. Arithmetic[1] and arithmetic[1] Sets and Relations.

For any formula $F(v_1)$ with v_1 as the only free variable, we say that $F(v_1)$ *expresses* the set of all numbers n such that $F(\overline{n})$ is a true sentence. Thus, $F(v_1)$ expresses the set A iff for all numbers n:

$$F(\overline{n}) \text{ is true} \leftrightarrow n \in A.$$

A regular formula $F(v_1, \ldots, v_n)$ will be said to *express* the set of all n-tuples $(k_1, \ldots, k_n)$ of natural numbers such that $F(\overline{k}_1, \ldots, \overline{k}_n)$ is a true sentence; so $F(v_1, \ldots, v_n)$ expresses the relation $R(x_1, \ldots, x_n)$ iff for all numbers $k_1, \ldots, k_n$, the following condition holds:

$$F(\overline{k}_1, \ldots, \overline{k}_n) \text{ is true} \leftrightarrow R(k_1, \ldots, k_n).$$

As an example, the set of even numbers is expressed by the formula $\exists v_2(v_1 = 0'' \cdot v_2)$ (since a number is even iff it is divisible by 2).

A set or relation will be called *Arithmetic* (note the capital "A") if it is expressed by some formula of $\mathcal{L}_E$. A set or relation is called *arithmetic* (note the small "a") if it is expressed by some formula of $\mathcal{L}_E$ in which the exponential symbol "E" does not occur. Arithmetic relations (and sets) can be informally characterized as those definable in first-order logic from plus, times and power; *arithmetic* relations are those definable from plus and times alone. [I did not include the less than or equal to relation because the relation $x_1 \leq x_2$ is itself definable from plus and times—in fact, just from plus—since it is expressed by the formula $\exists v_3(v_1 + v_3 = v_2)$.]

In a later chapter, we will prove a non-trivial result due to Gödel—namely that the exponential relation $x^y = z$ is itself definable from plus and times alone, and hence that every Arithmetic set and relation is arithmetic. But until this is proved, I will continue to use the

---

[1] The accent is on the syllable "met".

term "Arithmetic."

A function $f(x_1,\ldots,x_n)$ (from n-tuples of natural numbers to natural numbers) will be called Arithmetic if the relation

$$f(x_1,\ldots,x_n) = y$$

is Arithmetic. Thus, $f(x_1,\ldots,x_n)$ is Arithmetic iff there is a formula $F(v_1,\ldots,v_n,v_{n+1})$ such that for all numbers $x_1,\ldots,x_n$ and y, the sentence $F(\overline{x}_1,\ldots,\overline{x}_n,\overline{y})$ is true iff $f(x_1,\ldots,x_n) = y$.

We will sometimes say that a *property* P (of natural numbers) is Arithmetic, meaning that the set of numbers having the property P is Arithmetic. We will also use the collective term "condition" to mean either a relation or a property.

Exercise 2.

1. Show that the relation "x divides y" is arithmetic.
2. Show that the set of prime numbers is arithmetic.

Exercise 3. For any set A of natural numbers and any function $f(x)$ (from natural numbers to natural numbers) by $f^{-1}(A)$, we mean the set of all n such that $f(n) \in A$. Prove that if A and f are Arithmetic, then so is $f^{-1}(A)$. Show the same for "arithmetic."

Exercise 4.

1. Given two Arithmetic functions $f(x)$ and $g(y)$, show that the function $f(g(y))$ is Arithmetic.
2. Given two Arithmetic functions $f(x)$ and $g(x,y)$, show that the functions $g(f(y),y), g(x,f(y))$ and $f(g(x,y))$ are all Arithmetic.

Exercise 5. Let A be an infinite Arithmetic set. Then for any number y (whether in A or not), there must be an element x of A which is greater than y. Let $R(x,y)$ be the relation: x is the smallest element of A greater than y. Prove that the relation $R(x,y)$ is Arithmetic.

II. *Concatenation and Gödel Numbering*

§4. Concatenation to the Base b.

For any number $b \geq 2$, we will define a certain function $x *_b y$ called *concatenation to the base b* which will play a basic rôle throughout this volume.

We shall first give the definition for the familiar base 10. For any numbers m and n, we define $m *_{10} n$ to be the number which when

II. Concatenation and Gödel Numbering

written in ordinary base 10 notation consists of m followed by n. For example, $53 *_{10} 792 = 53792$. We note that $53792 = 53000 + 792$, which is $53 \cdot 10^3 + 792$. We note that 3 is the number of digits of 792 (when written in base 10 notation), or as we will say, 3 is the *length* of 792 (in base 10 notation).

More generally, $m *_{10} n = m \cdot 10^{\ell(n)} + n$, where $\ell(n)$ is the length of n (written in base 10 notation).

Still more generally, for any number $b \geq 2$, we define $m *_b n$ to be the number which, in base b notation, consists of m in base b notation followed by n in base b notation. Then $m *_b n = m \cdot b^{\ell_b(n)} + n$, where $\ell_b(n)$ is the length (number of base b digits) of n written in base b notation. The following proposition is basic.

Proposition 1. *For each $b \geq 2$, the relation $x *_b y = z$ is Arithmetic.*

Before turning to the proof, we state the essential idea. Consider first the familiar base 10. For any positive number n, the number $\ell_{10}(n)$ is simply the smallest number k such that $10^k > n$, and $10^{\ell_{10}(n)}$ is simply the smallest power of 10 greater than n (e.g.

$$10^{\ell_{10}(5368)} = 10^4 = 10,000,$$

which is the smallest power of 10 greater than 5368). In general, for any base $b \geq 2$ and for any number n, $b^{\ell_b(n)}$ is the smallest power of b greater than n if n is positive; otherwise b.

Proof of Proposition 1. Let b be any number ≥ 2.

1. Let $Pow_b(x)$ be the condition that x is a power of b. This condition is Arithmetic, because $Pow_b(x)$ holds iff $\exists y(x = b^y)$. [More formally, the set of powers of b is expressed by the formula $\exists v_2(v_1 = (\overline{b}\ \mathbf{E}\ v_2))$. We shall not be this formal in the future!]
2. The relation $b^{\ell_b(x)} = y$ (as a relation between x and y) is, as we have noted, equivalent to the condition that

$$(x = 0 \land y = b) \lor (x \neq 0 \land s(x, y)),$$

where $s(x, y)$ is the relation "y is the smallest power of b greater than x". It is Arithmetic because $s(x, y)$ holds iff

$$Pow_b(y) \land x < y \land \forall z((Pow_b(z) \land x < z) \supset y \leq z).$$

Hence it is Arithmetic. [The condition $x < y$ is equivalent to $x \leq y \land \sim (x = y)$. Hence it is Arithmetic.]
3. Finally, the relation $x \cdot b^{\ell_b(y)} + y = z$ (which is $x *_b y = z$) is the

condition:
$$\exists z_1 \exists z_2 (b^{\ell_b(y)} = z_1 \wedge x \cdot z_1 = z_2 \wedge z_2 + y = z).$$

Thus, the relation $x *_b y = z$ is Arithmetic.

Let us note that for any *positive* integers $x, y,$ and z,
$$(x *_b y) *_b z = x *_b (y *_b z).$$
But if $y = 0$, this can fail. [To use Quine's example,
$$(5 *_{10} 0) *_{10} 3 = 50 *_{10} 3 = 503,$$
but $5 *_{10} (0 *_{10} 3) = 5 *_{10} 3$ (since $0 *_{10} 3 = 3$), which is 53.] And so if we omit parentheses, it will be understood that parentheses are to be associated to the left (e.g. $x *_b y *_b z$ means $(x *_b y) *_b z$; not $x *_b (y *_b z)$)

Corollary 1. *For each $n \geq 2$ (and any $b \geq 2$), the relation*
$$x_1 *_b x_2 *_b \ldots *_b x_n = y$$
(as an $(n+1)$-ary relation among $x_1, \ldots, x_n, y$) is Arithmetic.

Proof. By induction on $n \geq 2$. We have already proved this for $n = 2$. Suppose n is a number ≥ 2 such that the relation $x_1 *_b \ldots *_b x_n = y$ is Arithmetic. Now, $x_1, *_b \ldots *_b x_n *_b x_{n+1} = y$ iff
$$\exists z (x_1 *_b \ldots *_b x_n = z \wedge z *_b x_{n+1} = y).$$
Hence this relation is also Arithmetic.

§5. Gödel Numbering. Arithmetic sentences (sentences of $\mathcal{L}_E$, that is) talk about *numbers*, not expressions of $\mathcal{L}_E$. The purpose of assigning Gödel numbers to expressions is to enable sentences to talk about expressions indirectly by directly talking about their Gödel numbers.

The Gödel numbering we will use is a modification of one due to Quine [1940]. Quine formulated his language in a 9-sign alphabet of symbols $S_1, S_2, \ldots, S_9$. Then to any compound expression $S_{i_1} S_{i_2} \ldots S_{i_n}$, he assigned the Gödel number which when written in base 10 notation is $i_1 i_2 \ldots i_n$ (e.g. the expression $S_3 S_1 S_2$ is assigned the number 312).

Our language $\mathcal{L}_E$ uses thirteen symbols, and so we will use Quine's idea, only taking concatenation to the base thirteen rather than the usual base ten. [The number thirteen happens to be a prime number,

II. Concatenation and Gödel Numbering

and we will see in a later chapter that taking concatenation to a prime base has certain technical advantages over a composite base like 10.]

We shall use "η", "ε" and "δ", as the base thirteen digits for 10, 11 and 12 respectively. We assign Gödel numbers to our thirteen symbols as follows:

$$\begin{array}{ccccccccccccc} 0 & ' & (&) & f & , & v & \sim & \supset & \forall & = & \leq & \sharp \\ 1 & 0 & 2 & 3 & 4 & 5 & 6 & 7 & 8 & 9 & \eta & \varepsilon & \delta \end{array}$$

Then, in any string of these symbols, we replace each symbol by its corresponding base 13 digit and read the resulting string of base 13 digits to the base 13. For example, the Gödel number of the string consisting of the third symbol followed by the sixth symbol followed by the second symbol is the number whose base 13 representation is "362"—i.e. the number $2 + (6 \cdot 13) + (3 \cdot 13^2)$.

For any $n > 0$, we let E_n be the expression whose Gödel number is n. Now, any string of accents has Gödel number 0, and we shall define E_0 to be just the accent symbol standing alone. We are, thus, using the word *expression* to mean any string of our 13 symbols that does not begin with an accent unless it is the accent by itself.

For any expressions E_x and E_y, by $E_x E_y$ we mean the expression consisting of E_x followed by E_y. It is immediate from our Gödel numbering rule that the Gödel number of $E_x E_y$ is $x *_{13} y$.

Our reason for choosing 0 as the Gödel number of the accent (prime) symbol is this: for any number n, the numeral $\bar{n}$ (like every other expression) has a Gödel number. We want our Gödel numbering to be such that the Gödel number of the numeral $\bar{n}$ is an Arithmetic function of n. Well, the numeral $\bar{n}$ consists of the symbol "0" followed by n accents, so its Gödel number, when written in base 13 notation, consists of "1" followed by n occurrences of "0". Hence it is simply 13^n.

Discussion. Actually, all results of this chapter and the next two go through for any Gödel numbering having the following two properties:

1. There is an Arithmetic function $x \circ y$ such that for any expressions X and Y with respective Gödel numbers x and y, the number $x \circ y$ is the Gödel number of XY;
2. For any n, the Gödel number of the numeral $\bar{n}$ is an Arithmetic function of n.

We chose the particular Gödel numbering we did primarily to

achieve these two properties as quickly as possible. We might also remark that any Gödel numbering possessing the first of the two properties must also possess the second (this will not be evident until a couple of chapters later; it rests on the particular way we chose to name the natural numbers).

Of course we *could* have used a base 10 instead of a base 13 Gödel numbering, say, by assigning as Gödel numbers to our thirteen symbols the numbers 1, 0, 2, 3, 4, 5, 6, 7, 89, 899, 8999, 89999, 899999 and 8999999, and the reader who feels more at home with the familiar base 10 notation may use this Gödel numbering if he wishes. Under this Gödel numbering, the Gödel number of $\bar{n}$ would be 10^n rather than 13^n. But, as we remarked before, there are certain technical advantages to working with a prime base like 13.

For the remainder of this chapter (and for all of the next chapters), I will be considering only the base 13 and, henceforth, write $x * y$ to mean $x *_{13} y$. [But the reader who prefers the base 10 can read "$x * y$" as $x *_{10} y$. But only whenever we write 13^x, he or she must write 10^x.]

III. Tarski's Theorem

§6. Diagonalization and Gödel Sentences.

We let T be the set of Gödel numbers of the *true* sentences of $\mathcal{L}_E$. This set T is a perfectly well defined set of natural numbers. Is it Arithmetic? We are going to show that it isn't (Tarski's theorem).

As in Chapter I, a sentence X will be called a *Gödel sentence* for a number set A if either X is true and its Gödel number is in A or X is false and its Gödel number is not in A. We now aim to show that for every Arithmetic set A, there is a Gödel sentence (from which Tarski's theorem will easily follow).

Almost a whole book could be written on various clever known methods for constructing Gödel sentences. Gödel's original method involves showing the existence of an Arithmetic function $\text{sub}(x, y)$ such that for any numbers x and y, if x is the Gödel number of a formula $F(v_1)$, then $\text{sub}(x, y)$ is the Gödel number of $F(\bar{y})$. To carry this out involves arithmetizing the operation of substituting numerals for variables, and this is a relatively complicated affair. Instead, we shall utilize a simple but clever idea due to Tarski [1953].

Informally, to say that a given property P holds for a given num-

III. Tarski's Theorem

ber n is equivalent to saying that for every number x equal to n, P holds for x. Formally, given any formula $F(v_1)$ with v_1 as the only free variable, the sentence $F(\bar{n})$ is equivalent to the sentence $\forall v_1(v_1 = \bar{n} \supset F(v_1))$. [Incidentally, it is also equivalent to the sentence $\exists v_1(v_1 = \bar{n} \wedge F(v_1))$.] The whole point now is that it is a relatively easy matter to show that the Gödel number of the sentence $\forall v_1(v_1 = \bar{n} \supset F(v_1))$ is an Arithmetic function of the Gödel number of $F(v_1)$ and the number n.

For any formula $F(v_1)$ and any n, we henceforth write $F[\bar{n}]$ to mean the sentence $\forall v_1(v_1 = \bar{n} \supset F(v_1))$. To repeat an important point, the sentences $F(\bar{n})$ and $F[\bar{n}]$, though not the same, are *equivalent* sentences (both true or both false.)

As a matter of fact, for *any* expression E, whether a formula or not, the expression $\forall v_1(v_1 = \bar{n} \supset E)$ is a perfectly well-defined expression (though a meaningless one if E is not a formula), and we shall write $E[\bar{n}]$ as an abbreviation of the (possibly meaningless) expression $\forall v_1(v_1 = \bar{n} \supset E)$. If E is a formula, then $E[\bar{n}]$ is also a formula but not necessarily a sentence. If E is a formula with v_1 as the only free variable, then $E[\bar{n}]$ is, of course, a sentence, but in all cases, $E[\bar{n}]$ is a well defined expression.

For any numbers e and n, by $r(e, n)$, we shall mean the Gödel number of the expression $E[\bar{n}]$ where E is the expression whose Gödel number is e. Thus, for any numbers x and y, the number $r(x, y)$ is the Gödel number of $E_x[\bar{y}]$. The function $r(x, y)$ plays a key rôle in this volume, and we now show that it is Arithmetic.

$E_x[\bar{y}]$ is the expression $\forall v_1(v_1 = \bar{y} \supset E_x)$. Let k be the Gödel number of the expression "$\forall v_1(v_1 =$". [The reader can write this number down if he wishes.] The implication symbol "$\supset$" has Gödel number 8 and the right parenthesis has Gödel number 3. The numeral $\bar{y}$ has Gödel number 13^y and the expression E_x has Gödel number x. Let us visualize the situation as

$$\underbrace{\forall v_1(v_1 =}_{k} \underbrace{\bar{y}}_{13^y} \underbrace{\supset}_{8} \underbrace{E_x}_{x} \underbrace{)}_{3}$$

We see that $E_x[\bar{y}]$ has Gödel number $k * 13^y * 8 * x * 3$ and so $r(x, y) = k * 13^y * 8 * x * 3$. The relation $r(x, y) = z$ is obviously Arithmetic (it can be written as $\exists w(w = 13^y \wedge z = k * w * 8 * x * 3)$).

We have, thus, proved

Proposition 2. *The function $r(x, y)$ is Arithmetic.*

The function $r(x,y)$ will crop up many times in this volume. We call it the *representation* function of $\mathcal{L}_E$.

Diagonalization. We now let $d(x) = r(x,x)$, and we call $d(x)$ the *diagonal* function. Since $r(x,y)$ is Arithmetic, then $d(x)$ is obviously Arithmetic. For any number n, $d(n)$ is the Gödel number of $E_n[\overline{n}]$.

For any number set A, we let A^* be the set of all n such that $d(n) \in A$. [Thus, $A^* = d^{-1}(A)$.]

Lemma 1. *If A is Arithmetic, then so is A^*.*

Proof. A^* is the set of all numbers x such that $\exists y(d(x) = y \wedge y \in A)$. Since the diagonal function $d(x)$ is Arithmetic, there is a formula $D(v_1, v_2)$ expressing the relation $d(x) = y$. Now suppose $F(v_1)$ is a formula expressing the set A. Then A^* is expressed by the formula $\exists v_2(D(v_1, v_2) \wedge F(v_2))$. [Alternatively, it is expressed by the formula $\forall v_2(D(v_1, v_2) \supset F(v_2))$.]

Theorem 1. *For every Arithmetic set A, there is a Gödel sentence for A.*

Proof. This really follows from the above lemma and Lemma D of Chapter 1 (cf. remarks below), but to repeat the proof for the particular language $\mathcal{L}_E$, suppose A is Arithmetic. Then A^* is Arithmetic by the above lemma. Let $H(v_1)$ be a formula expressing A^*, and let h be its Gödel number. Then $H[\overline{h}]$ is true $\leftrightarrow h \in A^* \leftrightarrow d(h) \in A$. But $d(h)$ is the Gödel number of $H[\overline{h}]$, and so $H[\overline{h}]$ is a Gödel sentence for A.

Remarks. Applying the abstract framework of Chapter 1 to our present language $\mathcal{L}_E$, we recall that we considered a function which assigned to every expression E and every number n, an expression $E(n)$. Well, for the language $\mathcal{L}_E$, we take $E(n)$ to be $E[\overline{n}]$, as we have defined it. Our "predicates" will now be formulas $F(v_1)$ in which v_1 is the only free variable. Then the lemma to Theorem 1 says that condition G_1 on page 7 holds for the language $\mathcal{L}_E$. Thus, Theorem 1 above is but a special case of (*b*) of Lemma D on page 8.

Tarski's Theorem. The class of Arithmetic sets is obviously closed under complementation because if $F(v_1)$ expresses A, then its negation $\sim F(v_1)$ expresses the complement $\tilde{A}$ of A. And so conditions G_1 and G_2 of Chapter 1 both hold for the language $\mathcal{L}_E$. Therefore, by Theorem T of Chapter 1, the set T of Gödel numbers of the true sentences of $\mathcal{L}_E$ is not expressible in $\mathcal{L}_E$—i.e. it is not Arithmetic.

To repeat the proof for the special case of $\mathcal{L}_E$, there cannot be a Gödel sentence for $\tilde{T}$, since such a sentence would be true if and only if it were not true. But if $\tilde{T}$ were arithmetic, there would be a Gödel sentence for $\tilde{T}$ by Theorem 1. Therefore, the set $\tilde{T}$ is not Arithmetic. Hence the set T is not Arithmetic either. And so we have proved

Theorem 2—Tarski's Theorem. *The set T of Gödel numbers of the true Arithmetic sentences is not Arithmetic.*

In the next chapter, we will turn to a formal axiom system for Arithmetic. On the surface, it will appear plausible that all true sentences are provable in the system, but it will turn out that the set of Gödel numbers of the *provable* sentences of the system, unlike the set T, *is* Arithmetic. Therefore, by Theorem 2, truth and provability don't coincide. In fact, using Th. 1, we will be able to exhibit a true sentence not provable in the system.

Exercise 6.

1. By the method we have just studied but using a base 10 Gödel numbering, find a Gödel sentence X for the set of even numbers. Then X is true just in case the Gödel number of X is even. Is the sentence X true or false?
2. Do the same for the set of odd numbers.

Exercise 7. Find an Arithmetic function $f(x)$ such that for any number n, if n is the Gödel number of a formula $F(v_1)$ with just the free variable v_1, $f(n)$ is the Gödel number of a Gödel sentence for the set expressed by $F(v_1)$. [Such a function $f(x)$ might aptly be called a *Gödelizer*].

Exercise 8. [Not too easy!] Prove that for any Arithmetic sets A and B there are sentences X and Y such that X is true iff A contains the Gödel number of Y, and Y is true iff B contains the Gödel number of X. This is an example of what might be thought of as *cross-reference*; X can be thought of as asserting that the Gödel number of Y is in A, and Y can be thought of as asserting that the Gödel number of X is in B.

Chapter III

The Incompleteness of Peano Arithmetic with Exponentiation

I. The Axiom System P.E.

§1. The Axiom System P.E. We shall now turn to a formal axiom system which we call *Peano Arithmetic with Exponentiation* and which we abbreviate "P.E.". We take certain correct formulas which we call *axioms* and provide two inference rules that enable us to prove new correct formulas from correct formulas already proved. The axioms will be infinite in number, but each axiom will be of one of nineteen easily recognizable forms; these forms are called *axiom schemes*. It will be convenient to classify these nineteen axiom schemes into four groups (cf. discussion that follows the display of the schemes). The axioms of Groups I and II are the so-called *logical axioms* and constitute a neat formalization of first-order logic with identity due to Kalish and Montague [1965], which is based on an earlier system due to Tarski [1965]. The axioms of Groups III and IV are the so-called *arithmetic* axioms.

In displaying these axiom schemes, F, G and H are any formulas, v_i and v_j are any variables, and t is any term. For example, the first scheme L_1 means that for *any* formulas F and G, the formula $(F \supset (G \supset F))$ is to be taken as an axiom; axiom scheme L_4 means that for any variable v_i and any formulas F and G, the formula

$$(\forall v_i(F \supset G) \supset (\forall v_i F \supset \forall v_i G))$$

is to be taken as an axiom.

I. The Axiom System P.E.

Group I—Axiom Schemes for Propositional Logic

L_1 : $(F \supset (G \supset F))$
L_2 : $(F \supset (G \supset H)) \supset ((F \supset G) \supset (F \supset H))$
L_3 : $((\sim F \supset \sim G) \supset (G \supset F))$

Group II—Additional Axiom Schemes for First-Order Logic with Identity.

L_4: $(\forall v_i(F \supset G) \supset (\forall v_i F \supset \forall v_i G))$
L_5: $(F \supset \forall v_i F)$, provided v_i does not occur in F.
L_6: $\exists v_i(v_i = t)$, provided v_i does not occur in t.
L_7: $(v_i = t \supset (X_1 v_i X_2 \supset X_1 t X_2))$, where X_1 and X_2 are any expressions such that $X_1 v_i X_2$ is an *atomic* formula. [Alternatively, this scheme can be written as $(v_i = t \supset (Y_1 \supset Y_2))$, where Y_1 is any atomic formula, and Y_2 is obtained from Y_1 by replacing any *one* occurrence of v_i in Y_1 by the term t.]

Group III—Eleven Axiom Schemes Having Only One Axiom Apiece

N_1 : $(v_1' = v_2' \supset v_1 = v_2)$
N_2 : $\sim \bar{0} = v_1'$
N_3 : $(v_1 + \bar{0}) = v_1$
N_4 : $(v_1 + v_2') = (v_1 + v_2)'$
N_5 : $(v_1 \cdot \bar{0}) = \bar{0}$
N_6 : $(v_1 \cdot v_2') = ((v_1 \cdot v_2) + v_1)$
N_7 : $(v_1 \leq \bar{0} \equiv v_1 = \bar{0})$
N_8 : $(v_1 \leq v_2' \equiv (v_1 \leq v_2 \lor v_1 = v_2'))$
N_9 : $((v_1 \leq v_2) \lor (v_2 \leq v_1))$
N_{10} : $(v_1 \text{ E } \bar{0}) = \bar{0}'$
N_{11} : $(v_1 \text{ E } v_2') = ((v_1 \text{ E } v_2) \cdot v_1)$

Group IV. This consists of only one axiom scheme—the scheme of mathematical induction—but infinitely many axioms, one for each formula $F(v_1)$. In displaying this scheme, $F(v_1)$ is to be any formula at all (it may contain free variables other than v_1). By $F[v_1']$ we shall mean *any one* of the formulas

$$\forall v_i(v_i = v_1' \supset \forall v_1(v_1 = v_i \supset F))$$

where v_i is any variable that does not occur in F. [Each of these formulas is equivalent to the formula $F(v_1')$—viz. the result of substituting the term v_1' for all free occurrences of v_1 in $F(v_1)$.] Here is

the scheme.
$$N_{12} : (F[\overline{0}] \supset (\forall v_1(F(v_1) \supset F[v_1']) \supset \forall v_1 F(v_1)))$$

Discussion. The axioms of Group I constitute a standard system for propositional logic (cf. Church [1956]). The axioms of Group II (due to Montague and Kalish) have the technical advantage of being stated without appeal to the notion of substitution of terms for free occurrences of variables—indeed, even the notions of free and bound occurrences of variables are circumvented. Instead, we have only the simpler notion of *replacement*—i.e. substitution of a term for *one* occurrence of a variable which is used in axiom scheme L_7.

Inference Rules. The inference rules of the system P.E. are two standard ones.

Rule 1 [Modus Ponens]—From F, $(F \supset G)$ to infer G.
Rule 2 [Generalization]—From F to infer $\forall v_i F$.

By a *proof* in the system P.E., we mean a finite sequence of formulas such that each member of the sequence is either an axiom or is directly derivable from two earlier members by Rule 1 or is directly derivable from some earlier member by Rule 2. A formula F is called *provable* (in the system P.E.) if there is a proof whose last member is F—such a sequence is called a *proof of F*. A formula is called *refutable* in P.E. if its negation is provable in P.E.

II. *Arithmetization of the Axiom System*

We now aim to prove that the set of Gödel numbers of the provable formulas of P.E. (unlike the set T of the last chapter) *is* an Arithmetic set.

§2. Preliminaries. We already know that for any base $b \geq 2$, the relation $x *_b y = z$ is Arithmetic and for each $n \geq 2$, the relation $x_1 *_b x_2 *_b \ldots *_b x_n = y$ is Arithmetic (Proposition 1 and its Corollary, last chapter).

We shall say that a number x *begins* a number y in base b notation if the base b representation of x is an initial segment of the base b representation of y. [For example, in base 10 notation, 593 begins 59348; also 593 begins 593.] The number 0 doesn't begin any number except 0. [We *don't* say that 0 begins 59, even though $59 = 059$.]

II. Arithmetization of the Axiom System

We write "xB_by" for "x begins y in base b notation". We say that x *ends* y in base b notation (in symbols, xE_by) if x is a final segment of y. [For example, in base 10 notation, 348 ends 59348; also 48 ends 59348, and so does 59348 itself. Also, 0 ends 570, as does 70, as does 570.] We say that x is *part* of y (in base b notation)—in symbols, xP_by—if x ends some number that begins y. [For example, in base 10 notation, 93 is part of 59348; so is 934, so is 34. Also 0 is part of 5076, but 0 is not part of 576.] Concerning the relation "x begins y" (in base b notation), if 0 is not part of y, then the relation holds iff either $x = y$ or $x \neq 0$ and $x *_b z = y$ for some z. However, in the more general case (that 0 might be part of y), x begins y iff either $x = y$ or $x \neq 0$ and $(x \cdot w) *_b z = y$ for some z and some power w of b. [For example, in base 10 notation, 5 begins 5007 since the above condition holds for $z = 7$ and $w = 100$. Also, 5 begins 57 since the condition holds for $z = 7$ and $w = 1$.] Let us note that the w and z involved are necessarily both less than or equal to (in fact less than) y—which is important, not for this chapter, but for the next.

The relations "x ends y" and "x is part of y" are even simpler to describe, and we have the following equivalences.

$$xB_by \leftrightarrow x = y \vee (x \neq 0 \wedge (\exists z \leq y)(\exists w \leq y)(\text{Pow}_b(w) \wedge (x \cdot w) *_b z = y))$$
$$xE_by \leftrightarrow x = y \vee (\exists z \leq y)(z *_b x = y)$$
$$xP_by \leftrightarrow (\exists z \leq y)(zE_by \wedge x B_b z)$$

Thus, the relations xB_by, xE_by and xP_by are all Arithmetic. Also

$$x_1 *_b x_2 *_b \ldots *_b x_n P_b y \leftrightarrow (\exists z \leq y) x_1 *_b \ldots *_b x_n = z \wedge z P_b y.$$

We, thus, have

Proposition 1. *For any $b \geq 2$ and any $n \geq 2$, the following relations are Arithmetic.*

1. xB_by
2. xE_by
3. xP_by
4. $x_1 *_b \ldots *_b x_n P_b y$

Remarks. In the above equivalences, each time we wrote

"$(\exists z \leq y)$"

we could have more simply written "$\exists z$", and this would have been enough to show that the relations in question are Arithmetic. Our purpose in writing the more complex expressions will emerge in later chapters in which we will need to know that these relations are not only Arithmetic, but belong to the much more narrow class of relations known as *constructive* arithmetic relations.

For the remainder of this chapter, b will be 13, and to reduce clutter, we delete the subscript b and just write xBy, xEy or xPy. Also we, henceforth, write xy for $x *_{13} y$. [This should lead to no confusion with multiplication since we always write $x \cdot y$ for x times y.] We also write $x\widetilde{P}y$ for $\sim xPy$ and $x_1 x_2 \ldots x_n P y$ for

$$x_1 *_{13} x_2 *_{13} \ldots *_{13} x_n P y.$$

Finite Sequences. The reader may have wondered what in the world the symbol "$\sharp$" is doing in our language $\mathcal{L}_E$, since it was never used for anything in the last chapter! Well, we have reserved it for forming *formal* sequences of expressions in the other twelve symbols. For any such expressions $X_1, \ldots, X_n$, the expression $\sharp X_1 \sharp X_2 \sharp \ldots X_n \sharp$ serves as the formal counterpart of the n-tuple $(X_1, \ldots, X_n)$, and its Gödel number will be called a *sequence* number.

Explained otherwise, we let K_{11} be the set of numbers n such that δ (the base 13 digit for 12) does not occur in (the base 13 representation of) n. All expressions in which the symbol "$\sharp$" does not occur have their Gödel numbers in the set K_{11} (this includes all numerals, variables, terms and formulas—all the so-called *meaningful* expressions). To any finite sequence $(a_1, \ldots, a_n)$ of numbers in K_{11}, we assign the number $\delta a_1 \delta a_2 \delta \ldots \delta a_n \delta$ which we call the *sequence number* of the sequence $(a_1, \ldots, a_n)$. We call x a *sequence number* if x is the sequence number of some finite sequence of elements of K_{11}. We let $\text{Seq}(x)$ be the property that x is a sequence number. We let $x \in y$ be the relation—y is a sequence number of some sequence of which x is a member. [Thus, for any numbers $x_1, \ldots x_n$ in K_{11}, if $y = \delta x_1 \delta \ldots \delta x_n \delta$, then $x \in y$ iff x is one of the numbers $x_1, \ldots, x_n$.] We let $x \prec_z y$ be the relation—z is the sequence number of a sequence in which x and y are members such that the first occurrence of x in the sequence is earlier than the first occurrence of y in the sequence.

Proposition 2. *Each of the conditions* $\text{Seq}(y)$, $x \in y$ *and* $x \prec_z y$ *is Arithmetic.*

Proof.

1. $\text{Seq}\, x \leftrightarrow \delta B x \wedge \delta E x \wedge \delta \neq x \wedge \delta\delta \widetilde{P} x \wedge (\forall y \leq x)(\delta 0 y P x \supset \delta B y)$
2. $x \in y \leftrightarrow \text{Seq}\, y \wedge \delta x \delta P y \wedge \delta \widetilde{P} x$
3. $x \prec_z y \leftrightarrow x \in z \wedge y \in z \wedge (\exists w \leq z)(wBz \wedge x \in w \wedge \sim y \in w)$

II. Arithmetization of the Axiom System

We, henceforth, write $(\forall x \in y)(----)$ to abbreviate
$$\forall x(x \in y \supset (----))$$
and we write $(\exists x, y \underset{w}{\prec} z)(----)$ to abbreviate
$$\exists x\, \exists y(x \underset{w}{\prec} z \wedge y \underset{w}{\prec} z \wedge (----)).$$

Formation Sequences. Our definitions in the last chapter of *terms* and *formulas* were inductive; we gave rules for constructing new terms from old ones and new formulas from old ones. We must now replace these inductive definitions by explicit definitions.

For any expressions X, Y and Z, define $\mathcal{R}_t(X, Y, Z)$ iff Z is one of the expressions $(X+Y), (X \cdot Y), (X \mathbin{E} Y)$ or X'. [We might call $\mathcal{R}_t$ the *formation relation for terms*.] By a formation *sequence* for terms, we mean a finite sequence $X_1, \ldots, X_n$ of expressions such that for each member X_i of the sequence, either X_i is a variable or a numeral, or there are earlier members X_j and X_k $(j < i, k < i)$ such that $\mathcal{R}_t(X_j, X_k, X_i)$. Then we can explicitly define an expression X to be a *term* iff there exists a formation sequence for terms of which X is a member.

Similarly with formulas. Define $\mathcal{R}_f(X, Y, Z)$ to hold if Z is one of the expressions $\sim X$ or $(X \supset Y)$, or Z is the expression $\forall v_i X$ for some variable v_i. [$\mathcal{R}_f$ can be called the formation relation for *formulas*.] Then we define a sequence $X_1, \ldots, X_n$ to be a formation sequence for *formulas* if for each $i \leq n$, either X_i is an atomic formula or there are numbers $j < i$ and $k < i$ such that $\mathcal{R}_f(X_j, X_k, X_i)$. Then an expression X is a formula if and only if there is a formation sequence for formulas of which X is a member.

§3. Arithmetization of the Syntax of P.E.

We recall that we are using the notation "E_x" for the expression whose Gödel number is x.

For any sequence $E_{x_1}, E_{x_2}, \ldots, E_{x_n}$ of expressions where $x_1, \ldots, x_n$ are all in K_{11}, by the Gödel number of the sequence $(E_{x_1}, \ldots, E_{x_n})$, we mean the sequence number of the sequence $(x_1, \ldots, x_n)$ of numbers. [It is the Gödel number of the expression $\sharp E_{x_1} \sharp E_{x_2} \sharp \ldots \sharp E_{x_n} \sharp$.]

We will list a chain of conditions (sets and relations) leading up to the key ones—$P_E(x)$ (E_x is a provable formula of P.E.) and $R_E(x)$ (E_x is a refutable formula of P.E.) and show that each condition is Arithmetic. For purposes that will be relevant only in a later chapter, the reader should note that the only universal quantifiers we will introduce are of the form $(\forall x \leq y)$, where y is a variable or a numeral (such quantifiers are called *bounded* universal quantifiers).

For any numbers x and y, we shall refer to the Gödel numbers of $(E_x \supset E_y)$, $\sim E_x$, $(E_x + E_y)$, $(E_y \cdot E_y)(E_x \text{ E } E_y)$, E_x', $E_x = E_y$ and $E_x \leq E_y$ as x imp y, neg(x), x pl y, x tim y, x exp y, $s(x)$, x id y and x le y respectively. Of course, these eight functions are all Arithmetic (for example x imp $y = 2x8y3$; neg$(x) = 7x$ (i.e. $7 * x$)).

Now we list the conditions (underneath each, we show that it is Arithmetic).

1. Sb(x)—E_x is a string of subscripts:
$$(\forall y \leq x)(yPx \supset 5Py)$$

2. Var(x)—E_x is a variable:
$$(\exists y \leq x)(\text{Sb}(y) \wedge x = 26y3)$$

3. Num(x)—E_x is a numeral:
$$Pow_{13}(x)$$

4. $R_1(x, y, z)$—The relation $\mathcal{R}_t(E_x, E_y, E_z)$ holds:
$$z = x \text{ pl } y \vee z = x \text{ tim } y \vee z = x \exp y \vee z = s(x)$$

5. Seqt (x)—E_x is a formation sequence for terms:
$$\text{Seq}(x) \wedge (\forall y \in x)(\text{Var}(y) \vee \text{Num}(y) \vee (\exists z, w \prec_x y) R_1(z, w, y))$$

6. tm (x)—E_x is a term:
$$\exists y(\text{Seqt}(y) \wedge x \in y)$$

7. f$_0(x)$—E_x is an atomic formula:
$$(\exists y \leq x)(\exists z \leq x)(\text{tm}(y) \wedge \text{tm}(z) \wedge (x = y \text{ id } z \vee x = y \text{ le } z))$$

8. Gen(x, y)—$E_y = \forall w E_x$ for some variable w:
$$(\exists z \leq y)(\text{Var}(z) \wedge y = 9zx)$$

9. $R_2(x, y, z)$—$\mathcal{R}_f(E_x, E_y, E_z)$ holds:
$$z = x \text{ imp } y \vee z = \text{neg}(x) \vee \text{Gen}(x, z)$$

10. Seqf(x)—E_x is a formation sequence for formulas:
$$\text{Seq}(x) \wedge (\forall y \in x)(\text{f}_0(y) \vee (\exists z, w \prec_x y) R_2(z, w, y))$$

II. Arithmetization of the Axiom System

11. $\text{fm}(x)$—E_x is a formula:
$$\exists y(\text{Seq}^{\text{f}}(y) \wedge x \in y)$$

12. $A(x)$—E_x is an axiom of P.E.:

See Note below

13. M.P. (x, y, z)—E_z is derivable from E_x and E_y by Rule 1:
$$y = x \text{ imp } z$$

14. Der (x, y, z)—E_z is derivable from E_x and E_y by Rule 1, or is derivable from E_x by Rule 2:
$$\text{M.P.}(x, y, z) \vee \text{Gen}(x, z)$$

15. $\text{Pf}(x)$—E_x is a proof in P.E.:
$$\text{Seq}(x) \wedge (\forall y \in x)(A(y) \vee (\exists z, w \prec_x y)\text{Der}(z, w, y))$$

16. $P_E(x)$—E_x is provable in P.E.:
$$\exists y(\text{Pf}(y) \wedge x \in y)$$

17. $R_E(x)$—E_x is refutable in P.E.:
$$P_E(\text{neg}(x))$$

Note on the Axioms. To show that $A(x)$ is Arithmetic, we break it up into nineteen parts (one for each of the axiom schemes). For each $n \leq 7$, we let $L_n(x)$ be the condition that E_x is an axiom of scheme L_n, and for each $n \leq 12$, we let $N_n(x)$ be the condition that E_x is an axiom of scheme N_n. The verifications that each of these nineteen conditions is Arithmetic are fairly uniform, and we give only sample cases.

Consider first $L_1(x)$. Well, E_x is an axiom of L_1 iff there are *formulas* E_y and E_z such that $E_x = (E_y \supset (E_z \supset E_y))$, and so $L_1(x)$ is the following condition:
$$(\exists y \leq x)(\exists z \leq x)(\text{fm }(y) \wedge \text{fm }(z) \wedge x = y \text{ imp}(z \text{ imp } y).$$

The conditions $L_2(x)$ and $L_3(x)$ can be handled similarly; we leave this to the reader.

As for Group II, let us take $L_4(x)$ as a sample case. Let $\varphi(y, z, w)$ be the Gödel number of $\forall E_y((E_z \supset E_w) \supset (\forall E_y E_z \supset \forall E_y E_w))$. The function $\varphi(x, y, z)$ is easily seen to be Arithmetic. Then $L_4(x)$ holds iff there are numbers y, z and w (all $\leq x$) such that $\text{var}(y)$, $\text{fm}(z), \text{fm}(w)$ and $x = \varphi(y, z, w)$. [The reader can write this all down symbolically, if desired]. We leave the other cases of Group II to the reader (only it should be noted that in L_6, we used the existential quantifier for abbreviation, and so in unabbreviated notation, L_6 is the scheme: $\sim \forall v_i \sim v_i = t$).

Group III is trivial since each of the schemes N_1–N_{11} contains only one axiom apiece, and so for each $i \leq 12$, $N_i(x)$ is simply the condition $x = g_i$, where g_i is the Gödel number of the axiom N_i.

Group IV (Axiom Scheme N_{12}) consists of all the induction axioms. To show that $N_{12}(x)$ is an Arithmetic condition, one first verifies that the relation E_x is a formula and E_y is one of the formulas, $E_x[v_1']$ is an Arithmetic relation between x and y. We leave this to the reader. It then becomes obvious that $N_{12}(x)$ is Arithmetic.

Having shown that each of the conditions $L_1(x), \ldots, L_7(x)$ and $N_1(x) \ldots, N_{12}(x)$ is Arithmetic, we take $A(x)$ to be the disjunction of these 19 conditions, and so $A(x)$ is Arithmetic.

This completes our arithmetization of the syntax of P.E. and so we have:

Proposition 3. *All the conditions* (1)–(17) *above are Arithmetic.*

§4. **Gödel's Incompleteness Theorem for P.E.** We let P_E be the set of Gödel numbers of the provable formulas of P.E. and R_E be the set of Gödel numbers of the refutable formulas of P.E. We have shown that these two sets are Arithmetic. Let $P(v_1)$ and $R(v_1)$ be formulas that express them in $\mathcal{L}_E$. Then the formula $\sim P(v_1)$ expresses the complement $\widetilde{P_E}$ of P_E. By the lemma to Theorem 1 of the last chapter, we can find a formula $H(v_1)$ expressing the set $\widetilde{P_E}^*$. Then, by the proof of Theorem 1, Ch. 2, its diagonalization $H[\overline{h}]$ is a Gödel sentence for the set $\widetilde{P_E}$. Hence is true iff it is not provable in P.E. Since P.E. is a correct system, then $H[\overline{h}]$ must be true but not provable in P.E. Since the sentence $\sim H[\overline{h}]$ is false, it is not provable in P.E. either.

Alternatively, using the dual argument of Chapter 1, since the set R_E is Arithmetic, so is R_E^*, so we can take a formula $K(v_1)$ expressing R_E^*. Its diagonalization $K[\overline{k}]$ is then a Gödel sentence for R_E. Hence it is true iff it is *refutable* in P.E. Then $K[\overline{k}]$ must be false but not refutable in P.E. Hence its negation $\sim K[\overline{k}]$ is true but not provable in P.E. Thus, $K[\overline{k}]$ (like $H[\overline{h}]$) is neither provable nor refutable in P.E. Either way, we have proved

Theorem 1. *The axiom system P.E. is incomplete.*

II. Arithmetization of the Axiom System

Remarks. The above incompleteness proof is the simplest one we know. Part of the simplicity is due to the use of the Tarski truth set (in chapters 5 and 6 we consider two other incompleteness proofs that do not use the truth set); part is due to the fact that we let exponentiation in on the ground floor (in the next chapter we consider a more austere system in which this is not the case) and part of the simplicity is due to our having taken the Montague-Kalish axiomatization of first-order logic with identity. To obtain an incompleteness proof for a more standard axiomatization of arithmetic, one must arithmetize the operation of substituting terms for free occurrences of variables in formulas. In the exercises below, we indicate the key steps of an incompleteness proof for the more standard formalization.

Exercise 1. Let $\mathrm{Fr}(x,y)$ be the relation "E_x is a variable, E_y is a formula and E_x has at least one free occurrence in E_y". Show that this relation is Arithmetic. [Hint: For any variable w and any expression X, w has a free occurrence in X iff there exists a finite sequence of expressions such that X is a member of the sequence, and for any member Y of the sequence, either Y is an atomic formula and w is part of Y, Y is the negation of some earlier member of the sequence, $Y = Y_1 \supset F$ or $Y = F \supset Y_1$ for some earlier member Y_1 of the sequence and some formula F (not necessarily a member of the sequence), or Y is the universal quantification of some earlier member of the sequence with respect to some variable distinct from w.]

Exercise 2. Using Exercise 1, show that:

1. The set of Gödel numbers of *sentences* is arithmetic;
2. The set of Gödel numbers of the provable *sentences* of P.E. is arithmetic. [It is more usual to work with this set rather than the set of Gödel numbers of the provable *formulas*. But we wanted our first incompleteness proof to be as simple as possible.]

Exercise 3. Given any finite sequence $(a_1, b_1), (a_2, b_2), \ldots, (a_n, b_n)$ of ordered pairs of numbers in K_{11}, we assign the sequence number $\delta\delta a_1 \delta b_1 \delta\delta \ldots \delta\delta a_n \delta b_n \delta\delta$. We let $\mathrm{Seq}_2(x)$ be the condition that x is the sequence number of a sequence of ordered pairs of numbers of K_{11}. We let $(x,y) \in z$ be the condition—z is the sequence number of a sequence of ordered pairs of numbers in K_{11} and (x,y) is a member of the sequence. We let $(x_1, y_1) \underset{z}{\prec} (x_2, y_2)$ be the relation—z is the sequence number of a sequence in which (x_1, y_1) occurs before

(x_2, y_2). Show that these three conditions are Arithmetic. [This is important for the next exercise.]

Exercise 4. For any term or formula E and any variable w and any term t, let E_t^w (sometimes written $E_w(t)$) be the result of substituting t for all free occurrences of w in E. Note that the following conditions hold.

1. If E is a numeral or a variable distinct from w, then $E_t^w = E$, but if $E = w$, then $E_t^w = t$ (in other words $w_t^w = t$)
2. If E is a term $r + s$, $r \cdot s$, $r \,\mathbf{E}\, s$ or r', then E_t^w is, respectively, $r_t^w + s_t^w$, $r_t^w \cdot s_t^w$, $r_t^w \,\mathbf{E}\, s_t^w$ or $r_t^{w'}$.
3. If E is an atomic formula $r = s$ or $r \leq s$, then E_t^w is, respectively, $r_t^w = s_t^w$ or $r_t^w \leq s_t^w$.
4. If E is a formula $F \supset G$ or $\sim F$, then E_t^w is, respectively, $F_t^w \supset G_t^w$ or $\sim F_t^w$.
5. If E is a formula $\forall v F$ where v is a variable distinct from w, then $E_t^w = \forall v F_t^w$. But if E is $\forall w F$, then $E_t^w = E$.

Now let $\text{Sub}(E, w, t, F)$ be the relation "E is a term or formula, w is a variable, t is a term and $F = E_t^w$." Let $\text{sub}(x_1, x_2, x_3, x_4)$ be the relation (between natural numbers): $\text{Sub}(E_{x_1}, E_{x_2}, E_{x_3}, E_{x_4})$.

(a) Using facts (1)–(5) above, $\text{Sub}(E_1, w, t, E_2)$ holds iff there exists a finite sequence of ordered pairs of expressions such that (E_1, E_2) is a member of the sequence, and for any member (X_1, X_2) of the sequence, either —— or there exist earlier members (Y_1, Y_2) and (Z_1, Z_2) such that ——. [Fill in the two blanks correctly.]

(b) Using (a) and Exercise 3, show that this relation is Arithmetic: $\text{sub}(x_1, x_2, x_3, x_4)$.

Exercise 5. For any variables w and w_1 and any formula F, w is said to be *bound* by w_1 in F if there is a formula G such that w has at least one free occurrence in G and $\forall w_1 G$ is part of F. A term t is said to be *substitutable* for a variable w in a formula F if w is not bound in F by any variable w_1 that occurs in t. Let $M(x, y, z)$ be the relation "E_x is substitutable for E_y in E_z" and show that this relation is Arithmetic.

Exercise 6. Let P.E.' be the axiom system P.E. with the axiom schemes of Group II replaced by the axiom schemes of the following group—call this Group II':

II. Arithmetization of the Axiom System

L_4' : Same as L_4
L_5' : $\forall w F \supset F_t^w$, provided t is substitutable for w in F.
L_6' : (a) $v_1 = v_1$
(b) $v_1 = v_2 \supset (v_3 = v_4 \supset (v_1 + v_3 = v_2 + v_4))$
(c) $v_1 = v_2 \supset (v_3 = v_4 \supset (v_1 \cdot v_3 = v_2 \cdot v_4))$
(d) $v_1 = v_2 \supset (v_3 = v_4 \supset (v_1 \operatorname{E} v_3 = v_2 \operatorname{E} v_4))$
(e) $v_1 = v_2 \supset v_1' = v_2'$
(f) $v_1 = v_2 \supset (v_3 = v_4 \supset (v_1 = v_3 \supset v_2 = v_4))$
(g) $v_1 = v_2 \supset (v_3 = v_4 \supset (v_1 \leq v_3 \supset v_2 \leq v_4))$

We note that axiom scheme L_5' has infinitely many axioms (w is any variable, F, F_1 and F_2 are any formulas).

The set of provable formulas of P.E.' is the same as the set of provable formulas of P.E. (this follows by the Montague-Kalish result [1965] that the Group I or Group II-axiomatization of first-order logic is equivalent to the Group I or Group II'-axiomatization). But a proof in P.E. is not the same thing as a proof in P.E.'. We let Pf'(x) be the condition—x is the Gödel number of a proof in P.E.'.

Using Exercises 4 and 5, show that the set of Gödel numbers of the axioms of L_5' is Arithmetic. Then it is easy to show that the condition Pf'(x) is Arithmetic, and hence that P.E.' is incomplete.

Chapter IV

Arithmetic Without the Exponential

I. The Incompleteness of P.A.

§1. By an *arithmetic* term or formula, we mean a term or formula in which the exponential symbol **E** does not appear, and by an arithmetic relation (or set), we mean a relation (set) expressible by an arithmetic formula. By the axiom system P.A. (Peano Arithmetic), we mean the system P.E. with axiom schemes N_{10} and N_{11} deleted, and in the remaining axiom schemes, *terms* and *formulas* are understood to be arithmetic terms and formulas. The system P.A. is the more usual object of modern study (indeed, the system P.E. is rarely considered in the literature). We chose to give the incompleteness proof for P.E. first since it is the simpler. In this chapter, we will prove the incompleteness of P.A. and establish several other results that will be needed in later chapters.

The incompleteness of P.A. will easily follow from the incompleteness of P.E., once we show that the relation $x^y = z$ is not only Arithmetic but arithmetic (definable from plus and times alone). We will first have to show that certain other relations are arithmetic, and as we are at it, we will show stronger results about these relations that will be needed, not for the incompleteness proof of this chapter, but for several chapters that follow—we will sooner or later need to show that certain key relations are not only arithmetic, but belong to a much narrower class of relations, the Σ_1-relations, which we will shortly define. These relations are the same as those known as *recursively enumerable*. Before defining the Σ_1-relations, we turn to a still narrower class, the Σ_0-relations, that will play a key role in our later development of recursive function theory.

I. The Incompleteness of P.A.

§2. We now define the classes of Σ_0-formulas and Σ_0-relations and then the Σ_1-formulas and relations.

By an *atomic* Σ_0-formula, we shall mean a formula of one of the four forms $c_1 + c_2 = c_3$, $c_1 \cdot c_2 = c_3$, $c_1 = c_2$ or $c_1 \leq c_2$, where each of c_1, c_2 or c_3 is either a variable or a numeral (but some may be variables and others numerals).

We now define the class of Σ_0-formulas by the following inductive scheme.

1. Every atomic Σ_0-formula is Σ_0.
2. If F and G are Σ_0, then so are $\sim F$ and $F \supset G$ (and hence, so are $F \wedge G$, $F \vee G$ and $F \equiv G$).
3. For any Σ_0-formula F, any variable v_i and every c which is either a numeral or a variable *distinct* from v_i, the expression $\forall v_i(v_i \leq c \supset F)$ is a Σ_0-formula.

We recall that $(\forall v_i \leq c)F$ is an abbreviation for $\forall v_i(v_i \leq c \supset F)$, and so if F is Σ_0, then so is $(\forall v_i \leq c)F$ (if c is any numeral or variable distinct from v_i).

We are also using $(\exists v_i \leq c)F$ to abbreviate $\sim(\forall v_i \leq c)\sim F$. Now, if F is Σ_0, then so is $\sim F$ (by (2)). Hence so is $(\forall v_i \leq c)\sim F$ (by (3), assuming $c \neq v_i$), and $\sim(\forall v_i \leq c)\sim F$ is Σ_0—i.e., $(\exists v_i \leq c)F$. [We note that $(\exists v_i \leq c)F$ is equivalent to $\exists v_i(v_i \leq c \wedge F)$].

The quantifiers $(\forall v_i \leq c)$ and $(\exists v_i \leq c)$ are sometimes called *bounded* quantifiers. Thus in a Σ_0-formula, all quantifiers are bounded.

A relation is called Σ_0 iff it is expressible by a Σ_0-formula. Σ_0-relations are also called *constructive arithmetic* relations.

Discussion. Let us informally observe that given any Σ_0-sentence (by which we mean a Σ_0-formula with no free variables), we can effectively decide whether it is true or false. This is certainly obvious for atomic Σ_0-sentences (obvious, that is, to the reader who knows how to add and multiply). Also, given any sentences X and Y, if we know how to determine the truth values of X and Y, we can obviously determine the truth values of $\sim X$ and of $X \supset Y$. Now let us consider the quantifiers. Suppose we have a formula $F(v_i)$ such that for every particular n, we can determine whether $F(\overline{n})$ is true or false. Can we then determine the truth value of the sentence $\exists v_i F(v_i)$? Not necessarily; if it is true, we can sooner or later know it by systematically examining the sentences $F(\overline{0}), F(\overline{1}), F(\overline{2}), \ldots$, but if it is not true, our search will be endless. Now, let us consider a sentence $(\exists v_i \leq \overline{k})F(v_i)$, where $\overline{k}$ is any numeral—specifically, let us consider the sentence $(\exists v_1 \leq \overline{5})F(v_1)$. Can we determine whether this sentence is true or false? Why certainly; we need merely test the sentences $F(\overline{0})$, $F(\overline{1})$, $F(\overline{2})$, $F(\overline{3})$, $F(\overline{4})$ and $F(\overline{5})$. Similar remarks apply to the universal quantifier. We have no effective test for $\forall v_i F(v_i)$ (if it is false, we will sooner or later know it, but if it is true, our search will be endless). However, for any number k, we have an effective test for the truth value of the sentence $(\forall v_i \leq \overline{k})F(v_i)$ (still assuming that for each n, we have an effective test for $F(\overline{n})$). And so we see that there is an effective test to decide which Σ_0-sentences are true and which ones are false.

Σ_1-**Relations.** By a Σ_1-formula, we mean a formula of the form $\exists v_{n+1} F(v_1, \ldots, v_n, v_{n+1})$, where $F(v_1, \ldots, v_n, v_{n+1})$ is a Σ_0-formula. We define a relation (or set) to be Σ_1 if and only if it is expressible by a Σ_1-formula. Thus, $R(x_1, \ldots, x_n)$ is a Σ_1-relation iff there is a Σ_0-relation $S(x_1, \ldots, x_n, y)$ such that for all $x_1, \ldots, x_n$, the equivalence $R(x_1, \ldots, x_n) \leftrightarrow \exists y S(x_1, \ldots, x_n, y)$ holds. We note that a Σ_1-formula begins with one unbounded existential quantifier. All the other quantifiers in it are bounded.

Σ-**Formulas.** We inductively define the class of Σ-formulas by the following rules:

1. Every Σ_0-formula is a Σ-formula.
2. If F is a Σ-formula, then for any variable v_i, the expression $\exists v_i F$ is a Σ-formula.
3. If F is a Σ-formula, then for any distinct variables v_i and v_j, the formulas $(\exists v_i \leq v_j)F$ and $(\forall v_i \leq v_j)F$ are Σ-formulas and for any numeral n, the formulas $(\exists v_i \leq \overline{n})F$ and $(\forall v_i \leq \overline{n})F$ are Σ-formulas.
4. For any Σ-formulas F and G, the formulas $F \vee G$ and $F \wedge G$ are Σ-formulas. If F is a Σ_0-formula and G is a Σ-formula, the formula $F \supset G$ is a Σ-formula.

A Σ-formula may contain any number of unbounded existential quantifiers, but all universal quantifiers must be bounded. We will call a relation (or set) a Σ-relation if it is expressed by some Σ-formula. In Part II of this chapter, we will show that the Σ-relations are the same as the Σ_1-relations (both are the same as those known as *recursively enumerable*).

We now aim to show that the exponential relation $x^y = z$ is not only arithmetic but is Σ_1. To this end, we first build up a useful armory of Σ_0-relations.

We immediately note that the relation $x < y$ is Σ_0, since $x < y$ iff $x \leq y \wedge x \neq y$. Also for any Σ_0-relation $R(x, y, z_1, \ldots, z_n)$, the relation $(\forall x < y)R(x,y,z_1,\ldots,z_n)$ is Σ_0, since it can be written $(\forall x \leq y)(x \neq y \supset R(x,y,z_1,\ldots,z_n))$.

§3. Concatenation to a Prime Base.

We already know that for any $b \geq 2$, concatenation to the base b is Arithmetic. The exponential function came into the definition of $x *_b y = z$ precisely at the point where we defined $Pow_b(x)$ (x is a power of b). And now we use a clever idea due to John Myhill. For any *prime* number p, we can define $Pow_p(x)$ without recourse to the exponential; x is a power of p if and only if every proper divisor of x is divisible by p! Using this idea, we will easily show that for a *prime* number p, the relation $x *_p y = z$ is arithmetic—in fact, even Σ_0.

Lemma 1. For every prime number p, the following conditions are Σ_0.

1. x div y—x divides y.
2. $Pow_p(x)$—x is a power of p.
3. $y = p^{\ell_p(x)}$—y is the smallest positive power of p greater than x.

Proof.

1. x div $y \leftrightarrow (\exists z \leq y)(x \cdot z = y)$.
2. $Pow_p(x) \leftrightarrow (\forall z \leq x)\,((z \text{ div } x \wedge z \neq 1) \supset p \text{ div } z)$.
3. $y = p^{\ell_p(x)} \leftrightarrow (Pow_p(y) \wedge y > x \wedge y > 1) \wedge$
   $\quad (\forall z < y) \sim (Pow_p(z) \wedge z > x \wedge z > 1)$.

Proposition A. For any prime p, the relation $x *_p y = z$ is Σ_0.

Proof.

$x *_p y = z \leftrightarrow x \cdot p^{\ell_p(y)} + y = z$
$\quad \leftrightarrow (\exists w_1 \leq z)(\exists w_2 \leq z)(w_1 = p^{\ell_p(y)} \wedge w_2 = x \cdot w_1 \wedge w_2 + y = z)$.

The condition $w_1 = p^{\ell_p(y)}$ is Σ_0 by the preceding lemma.

Proposition B. For each prime number p, the following relations are all Σ_0.

1. The relations $xB_p y$, $xE_p y$ and $xP_p y$ (x begins y, x ends y and x is part of y, in base p notation).
2. For each $n \geq 2$, the relation $x_1 *_p \cdots *_p x_n = y$.
3. For each $n \geq 2$, the relation $x_1 *_p \cdots *_p x_n P_p y$.

Proof.

1. We recall from Chapter 3 that xB_py is equivalent to

$$x = y \vee (x \neq 0 \wedge (\exists z \leq y)(\exists w \leq y)(\text{Pow}_p(w) \wedge (x \cdot w) *_p z = y))$$

$$xE_py \leftrightarrow x = y \vee (\exists z \leq y)(z *_p x = y)$$

$$xP_py \leftrightarrow (\exists z \leq y)(zE_py \wedge xB_pz)$$

Clearly each of these conditions is Σ_0.

2. The proof is by induction on $n \geq 2$. We already know that the relation $x_1 *_p x_2 = y$ is Σ_0. Now suppose $n \geq 2$ is such that the relation $x_1 *_p \cdots *_p x_n = y$ is Σ_0. Then the relation $x_1 *_p \cdots *_p x_n *_p x_{n+1} = y$ is also Σ_0, for it can be written as

$$(\exists z \leq y)(x_1 *_p \cdots *_p x_n = z \wedge z *_p x_{n+1} = y).$$

3. The relation $x_1 *_p \cdots *_p x_n P_p y$ can be written as

$$(\exists z \leq y)(x_1 *_p \cdots *_p x_n = z \wedge z P_p y).$$

Since 13 is a prime number, then all the parts of Propositions A and B hold for $p = 13$. From this it easily follows that the sets P_E and R_E of the last chapter are not only Arithmetic but arithmetic for the following reasons. Since the sets and relations of Propositions A and B are Σ_0 (for $p = 13$), then, of course, they are arithmetic. Then all the conditions (sets and relations) of Propositions 1, 2 and 3 of the last chapter are arithmetic, since the exponential function was never used in showing them to be Arithmetic once the conditions $Pow_{13}(x)$ and $x *_{13} y = z$ were shown to be Arithmetic. But we now know that the conditions $Pow_{13}(x)$ and $x *_{13} y = z$ are arithmetic. Therefore, all the conditions of Propositions 1, 2 and 3 of the last chapter are arithmetic.

Indeed, all the conditions of Propositions 1, 2 and 3 of the last chapter are not only arithmetic, but are in fact Σ, since the unbounded universal quantifier was never used. In particular, the sets P_E and R_E are Σ. The fact that they are Σ will be crucial for the next chapter. For this chapter, the only important thing is that they are arithmetic.

Since P_E is arithmetic, so is $\widetilde{P_E}$. However, it is a far cry to conclude that the set $\widetilde{P_E}^*$ is arithmetic, and we need *this* set to get a Gödel sentence for $\widetilde{P_E}$! To pass from the arithmeticity of a set A to the arithmeticity of A^* involves using the relation $13^x = y$ (in order to get the diagonal function), and, hence, we will have to show that the relation $13^x = y$ is arithmetic. [This does *not* follow from the mere fact that the set of powers of 13 is arithmetic!]

We know of no simpler method of showing the relation $13^x = y$ to be arithmetic than of showing the more general fact that the relation $x^y = z$ is arithmetic (even though 13 is a prime number). [If there is a simpler method, we would love to know about it!] And so we now turn to the task of showing that the exponential relation is arithmetic—in fact, we want to prove:

Theorem E. The relation $x^y = z$ is Σ_1.

§4. The Finite Set Lemma. Theorem E will easily follow once we have proved the following lemma.

Lemma K. There is a constructive arithmetic relation $K(x, y, z)$ having the following two properties.

1. For any finite sequence $(a_1, b_1), (a_2, b_2), \ldots, (a_n, b_n)$ of ordered pairs of natural numbers, there is a number z such that for any numbers x and y, the relation $K(x, y, z)$ holds if and only if (x, y) is one of the pairs $(a_1, b_1), \ldots, (a_n, b_n)$.
2. For any numbers x, y and z, if $K(x, y, z)$ holds, then $x \leq z$ and $y \leq z$.

To prove Lemma K, we use Propositions A and B. These propositions hold for *any* prime number p—in particular, for p the number 13. [Actually, any other prime number would serve as well—e.g., $p = 2$—but since we are now used to the prime base 13, we will stick to it.] To avoid unnecessary circumlocution, until further notice, we will identify the natural numbers with their base 13 representations. Thus, e.g., we will say that a number x is part of a number y, meaning that the base 13 representation of x is part of the base 13 representation of y.

Now we use a clever idea due to Quine [1946]. By a *frame* we shall mean a number of the form $2t2$, where t is a string of 1's. We let $1(x)$ be the condition that x is a string of 1's (in base 13 notation, of course.) The condition $1(x)$ is Σ_0, for

$$1(x) \leftrightarrow x \neq 0 \land (\forall y \leq x)(yPx \supset 1Py).$$

Now let θ be any finite sequence $((a_1, b_1), \ldots, (a_n, b_n))$ of ordered pairs of numbers, and let f be any frame which is longer than any frame which is part of any of the numbers $a_1, b_1, \ldots, a_n, b_n$. For such an f, we call the number $ffa_1fb_1ff \cdots ffa_nfb_nff$ a *sequence num-*

ber of θ. [There is no need to assign to θ a *unique* sequence number. If desired, we could, of course, take the *least* sequence number, but this is an unncessary technical complication. We note that "sequence number" has a very different meaning in this chapter than in the last! If we were interested in assigning sequence numbers to only those θ such that $a_1, b_1, \ldots, a_n, b_n$ are all in K_{11}, then the obvious thing to do would be to take the number $\delta\delta a_1 \delta b_1 \delta\delta a_2 \delta b_2 \delta\delta \cdots \delta\delta a_n \delta b_n \delta\delta$. However, for our present purposes, we need to consider sequences of ordered pairs of *any* natural numbers whatsoever. The frame f now plays the rôle formally played by δ.]

We call x a *maximal frame* of y if x is a frame, x is part of y and x is as long as any frame that is part of y. Let x mf y be the relation that x is a maximal frame of y. The relation x mf y is Σ_0 as

$$x \text{ mf } y \leftrightarrow xPy \wedge (\exists z \leq y)(1(z) \wedge$$
$$x = 2z2 \wedge \sim (\exists w \leq y)(1(w) \wedge 2zw2Py)).$$

Now we can define the crucial Σ_0-relation $K(x, y, z)$.

$$K(x, y, z) \underset{\text{df}}{=} (\exists w \leq z)(w \text{ mf } z \wedge wwxwyww Pz \wedge w\widetilde{P}x \wedge w\widetilde{P}y).$$

We note that for any sequence θ of ordered pairs of numbers, if z is any sequence number of θ, then $K(x, y, z)$ holds if and only if the ordered pair (x, y) is a member of the sequence θ. It is obvious from the definition of $K(x, y, z)$ that for *any* numbers x, y and z, if $K(x, y, z)$ holds, then $x \leq z$ and $y \leq z$ (in fact, $x < z$ and $y < z$), and so we have proved Lemma K.

Remarks. An alternative method of proving Lemma K, which does not use Quine's maximal frames, uses an ingenious idea due to Saul Kripke. Not only is it the case that for each prime p the relation $x *_p y = z$ is Σ_0, but the 4-place relation "p is a prime and $x *_p y = z$" (as a relation among p, x, y, z) is easily seen to be Σ_0. Now, given any sequence θ of ordered pairs, take a prime number p such that $p - 1$ is greater than any of the numbers $a_1, b_1, \ldots, a_n, b_n$. Let $s = p - 1$. Then in base p notation, the numbers $s, a_1, b_1, \ldots, a_n$ and b_n are all single digits! We let q be the number $ssa_1 sb_1 ss \cdots ssa_n sb_n ss$. The number q serves perfectly as a sequence number of θ. [Details of this can be found in Boolos and Jeffrey [1980], Ch. 14, pp. 162, 163].

§5. Proof of Theorem E.

Now that we have Lemma K, the proof of Theorem E is easy.

We note that $x^y = z$ iff there exists a set S of ordered pairs such

I. The Incompleteness of P.A.

that:

(1) $(y, z) \in S$.
(2) For every pair (a, b) in S, either $(a, b) = (0, 1)$ or there is some pair (c, d) in S such that $(a, b) = (c + 1, d \cdot x)$

We can see this as follows: If $x^y = z$, then we can take S to be the set $\{(0, 1), (1, x), (2, x^2), \ldots, (y, x^y)\}$. Conversely, suppose S is any set of ordered pairs such that (1) and (2) hold. From (2) it follows that for any pair (a, b) in S, it must be the case that $x^a = b$ (as can be seen by induction on a). And so by (1), we have $x^y = z$.

It follows from this that $x^y = z$ iff there exists a number w such that $K(y, z, w)$, and for any number $a \leq w$ and any $b \leq w$, if $K(a, b, w)$, then either $a = 0$ and $b = 1$, or there are numbers $c \leq a$ and $d \leq b$ such that $K(c, d, w)$ and $a = c + 1$ and $b = d \cdot x$. Thus, $x^y = z$ iff the following condition holds:

$$\exists w(K(x, y, w) \wedge (\forall a \leq w)(\forall b \leq w)(K(a, b, w) \supset ((a = 0 \wedge b = 1) \vee$$
$$(\exists c \leq a)(\exists d \leq b)(K(c, d, w) \wedge a = c + 1 \wedge b = d \cdot x)))).$$

A variant of the above proof uses what is known as a *Beta-function*. We discuss this in the following section.

Beta-Functions. A function $\beta(x, y)$ is called a *Beta-function* if for every finite sequence $(a_0, a_1, \ldots, a_n)$ of numbers there is a number w such that $\beta(w, 0) = a_0$ and $\beta(w, 1) = a_1, \ldots, \beta(w, n) = a_n$.

We call a function $f(x_1, \ldots, x_n)$ a Σ_0-function (or a constructive arithmetic function) if the relation $f(x_1, \ldots, x_n) = y$ is Σ_0.

From Lemma K we easily obtain

Theorem B—The Beta-Function Theorem. There is a constructive arithmetic Beta-function.

Proof. Using the Σ_0-relation $K(x, y, z)$ of Lemma K, define $\beta(w, i)$ to be the smallest number k such that $K(i, k, w)$, if there is such a number k. Otherwise set $\beta(w, i) = 0$. The relation $\beta(w, x) = y$ is Σ_0 because it can be expressed as

$$(K(x, y, w) \wedge (\forall z < y) \sim K(x, z, w)) \vee (\sim (\exists z \leq w) K(x, z, w) \wedge y = 0).$$

Now, given any finite sequence $(a_0, \ldots, a_n)$, if we let w be a sequence number of the sequence $((0, a_0), (1, a_1), \ldots, (n, a_n))$, then for each $i \leq n$, $K(i, a_i, w)$ holds, and a_i is the only number m for which $K(i, m, w)$ holds, and so $\beta(w, i) = a_i$. This concludes the proof.

Remarks (1). The first β-function was constructed by Gödel [1931]. Its definition requires the result from number theory known as the *Chinese Remainder Theorem*. In T.F.S. we introduced a simplified β-function (much like the one above) that avoids the Chinese remainder theorem. Gödel's beta-function belongs to the class of functions known as *primitive recursive*, of which the Σ_0-functions form a small subclass. The β-function constructed in T.F.S. belongs to the still smaller class called *strictly rudimentary* functions, which are closely tied up with the theory of finite automata (cf. T.F.S., Chapter IV for details).

(2). Using a β-function, we have the following alternative proof of Theorem E.

Obviously $x^y = z$ iff there exists a finite sequence $(a_0, a_1, \ldots, a_y)$ (namely $(1, x, x^2, \ldots, x^y)$) such that $a_0 = 1$ and $a_y = z$ and for each $i < y$ and $a_{i+1} = a_i \cdot x$. And so $x^y = z$ iff

$$\exists w(\beta(w,0) = 1 \wedge \beta(w,y) = z \wedge (\forall n < y)(\beta(w,n+1) = \beta(w,n) \cdot x)).$$

Note. The condition $\beta(w, n+1) = \beta(w,n) \cdot x$ is Σ_0 and can be written as

$(\exists m_1 \leq w)(\exists m_2 \leq w)(\exists m_3 \leq w)(m_1 = n + 1 \wedge \beta(w, m_1)$
$= m_2 \wedge \beta(w,n) = m_3 \wedge m_2 = m_3 \cdot x).$

Theorem E has the following important corollaries:

Corollary 1. *For any arithmetic set A, the set A^* is arithmetic. If A is Σ, then so is A^*.*

Proof. Since the relation $x^y = z$ is Σ_1, it is certainly Σ. Hence the relation $13^x = y$ (as a relation between x and y) is Σ (in fact, Σ_1). Hence the diagonal function $d(x)$ is Σ (we recall that $d(x) = y$ iff $\exists z(z = 13^x \wedge kz8x3 = y)$, where k is the Gödel number of the expression "$\forall v_1(v_1 =$"). And so there is a Σ-formula $D(v_1, v_2)$ expressing the relation $d(x) = y$. Then, for any formula $A(v_1)$ expressing a set A, the formula $\exists v_2(D(v_1, v_2) \wedge A(v_2))$ expresses the set A^*. So if A is arithmetic, then so is A^*. If $A(v_1)$ happens to be a Σ-formula, then so is the formula $\exists v_2(D(v_1, v_2) \wedge A(v_2))$.

Corollary 2—Tarski's Theorem for $\mathcal{L}_A$. *The set of Gödel numbers of the true arithmetic sentences is not arithmetic.*

Proof. Let T_A be the set of Gödel numbers of the true *arithmetic* sentences. If T_A were arithmetic, then $\widetilde{T_A}$ would be arithmetic. Hence $\widetilde{T_A}^*$ would be arithmetic (by Corollary 1) and we would have the same contradiction as in the proof of Tarski's theorem for $\mathcal{L}_E$—namely, there would be an *arithmetic* formula $H(v_1)$ such that for any n, $H(\bar{n})$ would be true $\leftrightarrow n \in \widetilde{T_A}^*$. Hence $H[\bar{h}]$ would be true $\leftrightarrow h \in \widetilde{T_A}^* \leftrightarrow H[\bar{h}]$ is not true.

Corollary 3. The sets P_E^* and R_E^* are Σ. The set $\widetilde{P_E}^*$ is arithmetic.

Proof. We have already proved that the sets P_E and R_E are Σ. Hence the sets P_E^* and R_E^* are also Σ by Corollary 1.

Since the set P_E is Σ, it is arithmetic. Hence its complement $\widetilde{P_E}$ is arithmetic. Therefore, the set $\widetilde{P_E}^*$ is arithmetic by Corollary 1.

§6. The Incompleteness of Peano Arithmetic.

Since the set $\widetilde{P_E}^*$ is arithmetic, there is an *arithmetic* formula $H(v_1)$ that expresses $\widetilde{P_E}^*$; its diagonalization $H[\bar{h}]$ is then an *arithmetic* Gödel sentence for P_E, and is, thus, true but not provable in P.E. Since it is not provable in P.E., then it is certainly not provable in P.A. (whose set of axioms is a proper subset of the set of axioms of P.E.). And so $H[\bar{h}]$ is true but not provable in P.A. Since $\sim H[\bar{h}]$ is false, then $\sim H[\bar{h}]$ is also not provable in P.A., so $H[\bar{h}]$ is a sentence in the language $\mathcal{L}_A$ of P.A. which is neither provable nor refutable in the axiom system P.A.

Of course, we did not need to use the axiom system P.E. to prove the incompleteness of P.A. Let P_A be the set of Gödel numbers of the formulas provable in P.A. To show that the set P_A is arithmetic (in fact Σ)—a fact we will need for subsequent chapters—we need make only a couple of trivial changes in the proof of Proposition 3 of the last chapter. In item (4) on page 34, $R_1(x,y,z)$ should now be: E_z is one of the expressions $(E_x + E_y)$, $(E_x \cdot E_y)$ or E_x', so in the proof of (4), just delete the disjunctive clause $z = x \exp y$. With this change, $\text{tm}(x)$ and $\text{fm}(x)$ are then the conditions that E_x is a term or formula of P.A. rather than P.E. Then in $A(x)$ (item (12)), delete the disjunctive clauses $N_{10}(x)$ and $N_{11}(x)$. Then $A(x)$ is now the condition that E_x is an axiom of P.A. With these changes, items 16 and 17—which we now label $P_A(x)$ and $R_A(x)$—thus become: E_x is provable and refutable in P.A.

And so the set P_A is Σ and, hence, the sets $\widetilde{P_A}$ and $\widetilde{P_A}^*$ are arithmetic. If we now take an arithmetic formula $H(v_1)$ expressing the set $\widetilde{P_A}^*$, its diagonalization $H[\bar{h}]$ then expresses its own non-provability in the system P.A. rather than P.E. Either way, we have proved

Theorem I. The system P.A. is incomplete.

Exercise 1. Let G be the sentence $H[\bar{h}]$ which is true but not provable in P.A. ($H(v_1)$ is a formula that expresses the set $\widetilde{P^*}$). Suppose we add the sentence G as an axiom to P.A.—call this system P.A. $+ \{G\}$. Since G is a true sentence, the system P.A. $+ \{G\}$ is also a correct system. Is it complete?

II. More on Σ_1-Relations

For several chapters that follow, we will need to know that every Σ-relation (and set) is also Σ_1. This fact follows rather easily from the following proposition.

Proposition C.

(a) Every Σ_0-relation is also Σ_1.
(b) If $R(x_1,\ldots,x_n,y)$ is Σ_1, then so is the relation
$$\exists y R(x_1,\ldots,x_n,y).$$
(c) If $R_1(x_1,\ldots,x_n)$ and $R_2(x_1,\ldots,x_n)$ are Σ_1, then so are the relations
$$R_1(x_1,\ldots,x_n) \vee R_2(x_1,\ldots,x_n)$$
and
$$R_1(x_1,\ldots,x_n) \wedge R_2(x_1,\ldots,x_n).$$
(d) If $R(x_1,\ldots,x_n,y,z)$ is Σ_1, then so are the relations
$$(\exists y \leq z)R(x_1,\ldots,x_n,y,z)$$
and
$$(\forall y \leq z)R(x_1,\ldots,x_n,y,z).$$
(e) If R is Σ_0 and S is Σ_1, then the relation $R \supset S$ is Σ_1.

Proof.

(a) Suppose the relation $R(x_1,\ldots,x_n)$ is Σ_0. Let $F(v_1,\ldots,v_n)$ be a Σ_0-formula that expresses it. Then the formula
$$\exists v_{n+1} F(v_1,\ldots,v_n)$$

II. More on Σ_1-Relations

is a Σ_1-formula, and it expresses the same relation $R(x_1,\ldots,x_n)$ (vacuous quantification!). So R is Σ_1.

(b) Let us first note that for any relation $S(x_1,\ldots,x_n,y,z)$, the following two conditions are equivalent:

(1) $\exists y \exists z S(x_1,\ldots,x_n,y,z)$
(2) $\exists w (\exists y \leq w)(\exists z \leq w) S(x_1,\ldots,x_n,y,z)$.

It is obvious that (2) implies (1). Now suppose $x_1,\ldots,x_n$ are numbers such that (1) holds. Then there are numbers y and z such that $S(x_1,\ldots,x_n,y,z)$ holds. Let w be the maximum of y and z. Then $(\exists y \leq w)(\exists z \leq w)S(x_1,\ldots,x_n,y,z)$ holds for such a number w, and hence (2) holds.

Suppose now $R(x_1,\ldots,x_n,y)$ is Σ_1. Then $R(x_1,\ldots,x_n,y)$ is of the form $\exists z S(x_1,\ldots,x_n,y,z)$, where S is a Σ_0-relation. Then the relation $\exists y R(x_1,\ldots,x_n,y)$ is the same as the relation $\exists y \exists z S(x_1,\ldots,x_n,y,z)$. But as shown above, this is the same as the relation $\exists w(\exists y \leq w)(\exists z \leq w)S(x_1,\ldots,x_n,y,z)$, and this relation is Σ_1 because the relation

$$(\exists y \leq w)(\exists z \leq w)S(x_1,\ldots,x_n,y,z)$$

is a Σ_0-relation among $x_1,\ldots,x_n$ and w.

(c) This statement is equivalent to the statement that for any Σ_0-relations $S_1(x_1,\ldots,x_n,y)$ and $S_2(x_1,\ldots,x_n,y)$, the two relations $\exists y S_1(x_1,\ldots,x_n,y) \lor \exists y S_2(x_1,\ldots,x_n,y)$ and the relation $\exists y S_1(x_1,\ldots,x_n,y) \land \exists y S_2(x_1,\ldots,x_n,y)$ are both Σ_1. Well, the first is equivalent to

$$\exists y (S_1(x_1,\ldots,x_n,y) \lor S_2(x_1,\ldots,x_n,y));$$

the second is equivalent to

$$\exists y \exists z (S_1(x_1,\ldots,x_n,y) \land S_2(x_1,\ldots,x_n,z)),$$

which is Σ_1 by (b).

(d) Suppose $R(x_1,\ldots,x_n,y,z)$ is Σ_1. The proof that the relation $(\exists y \leq z)R(x_1,\ldots,x_n,y,z)$ is Σ_1 is pretty obvious. We know that the relation $y \leq z$ is Σ_0. Hence the set K of all $n+2$-tuples $(x_1,\ldots,x_n,y,z)$ such that $y \leq z$ is Σ_0 (it is expressed by the Σ_0-formula $v_1 = v_1 \land \cdots \land v_n = v_n \land v_{n+1} \leq v_{n+2}$). Hence the relation $K(x_1,\ldots,x_n,y,z)$ is Σ_1 (by (a)). Therefore, the relation $K(x_1,\ldots,x_n,y,z) \land R(x_1,\ldots,x_n,y,z)$ is Σ_1 (by (c)), but this is simply the relation $y \leq z \land R(x_1,\ldots,x_n,y,z)$. Then

by (b), the relation
$$\exists y(y \leq z \wedge R(x_1,\ldots,x_n,y,z))$$
(as a relation between $x_1,\ldots,x_n,z$) is Σ_1, but this is the relation $(\exists y \leq z)R(x_1,\ldots,x_n,y,z)$.

The proof that the relation $(\forall y \leq z)R(x_1,\ldots,x_n,y,z)$ is Σ_1 is more subtle and more interesting! Since R is Σ_1, there is a Σ_0-relation $S(x_1,\ldots,x_n,y,z,w)$ such that for all $x_1,\ldots,x_n,y$ and z, $R(x_1,\ldots,x_n,y,z)$ holds iff $\exists w S(x_1,\ldots,x_n,y,z,w)$. Then $(\forall y \leq z)R(x_1,\ldots,x_n,y,z)$ is the condition
$$(\forall y \leq z)\exists w S(x_1,\ldots,x_n,y,z,w).$$

Now suppose $x_1,\ldots,x_n$ and z are numbers such that
$$(\forall y \leq z)\exists w S(x_1,\ldots,x_n,y,z,w)$$
holds. Then for every $y \leq z$, there is some number w_y such that $S(x_1,\ldots,x_n,y,w_y)$ holds. Let v be the greatest of the numbers $w_0, w_1, \ldots, w_z$. Then $w_0,\ldots,w_z$ are all $\leq v$, and so for every $y \leq z$, there is some $w \leq v$ (namely w_y) such that $S(x_1,\ldots,x_n,y,z,w)$. Thus (for this v), the condition
$$(\forall y \leq z)(\exists w \leq v)S(x_1,\ldots,x_n,y,z,w)$$
holds and, therefore, the condition
$$\exists v(\forall y \leq z)(\exists w \leq v)S(x_1,\ldots,x_n,y,z,w)$$
holds. Conversely, this condition obviously implies the condition
$$(\forall y \leq z)\exists w S(x_1,\ldots,x_n,y,z,w),$$
which is the condition
$$(\forall y \leq z)R(x_1,\ldots,x_n,y,z).$$

(e) Suppose R is Σ_0 and S is Σ_1. Then $\widetilde{R}$ is also Σ_0. Hence $\widetilde{R}$ is Σ_1 by (a). Then $\widetilde{R} \vee S$ is Σ_1 by (c). But $\widetilde{R} \vee S$ is the relation $R \supset S$.

For any formula $F(v_{i_1},\ldots,v_{i_k})$ ($i_1 < i_2 < \ldots < i_k$) and any $n \geq i_k$, by $F^{(n)}$, we shall mean the set of all n-tuples $(a_1,\ldots,a_n)$ such that $F(\overline{a}_{i_1},\ldots,\overline{a}_{i_k})$ is a true sentence. (For example, if F is the formula $v_3 + v_1 = v_5$, then $F^{(6)}$ is the set of all sextuples $(a_1,a_2,a_3,a_4,a_5,a_6)$ such that $a_3+a_1 = a_5$.) If F is a *regular* formula $F(v_1,\ldots,v_n)$, then, of course, $F^{(n)}$ is simply the relation expressed by $F(v_1, \ldots, v_n)$.

By (a)–(e) of Proposition C, it follows by a straightforward induction argument on degrees of formulas that for any Σ-formula $F(v_{i_1}, \ldots, v_{i_k})$ and any $n \geq i_k$, the relation (or set) $F^{(n)}$ is Σ_1. It then follows that any *regular* Σ-formula $F(v_1, \ldots, v_n)$ expresses a Σ_1-relation, and so we have:

Proposition C$_1$. Every Σ-relation is Σ_1.

From Proposition C$_1$ and the corollaries of Theorem E, we have:

Corollary 1. If A is Σ_1, then so is A^*.

Corollary 2. The sets P_A^* and R_A^* are Σ_1.

Appendix

A set or relation is called *recursive* if it and its complement are both Σ_1. [The complement of a relation $R(x_1, \ldots, x_n)$ is the relation $\sim R(x_1, \ldots, x_n)$.] A function $f(x_1, \ldots, x_n)$ is called recursive if the relation $f(x_1, \ldots, x_n) = y$ is recursive.

Many of the sets and relations which were shown in the last chapter to be Arithmetic, and which were shown in this chapter to be Σ_1, are in fact recursive. Indeed, the relations (1)–(15) of Proposition 3, Ch. 3 (pp. 33–34) are all recursive. The standard proof of this uses a device known as "course of values recursion", but we have the following alternative method which is much simpler.

We define $\pi(x)$ to be $13^{(x^2+x+1)}$. The function $\pi(x)$ is recursive (Ex. 2 below). Its significance is given by the following theorem.

Theorem D. For any n and any $k \leq n$ and any sequence $(a_1, \ldots, a_k)$ of numbers of K_{11} all of which are $\leq n$, its sequence number

$$\delta a_1 \delta \ldots \delta a_k \delta$$

is $\leq \pi(n)$.

Proof. Let x be this number. Then $x \leq y$, where $y = \underbrace{\delta n \delta n \ldots \delta n \delta}_{n}$

(i.e. $\delta * n$ followed by itself n times followed again by δ). We show that $y \leq \pi(n)$.

For any number z, we let $L(z)$ be the length of z (in base 13 notation). Then $L(y) = n \cdot L(n) + n + 1$ (since δ is of length 1). Also $L(n) \leq n$, and so $L(y) \leq n^2 + n + 1$. Also $y \leq 13^{L(y)}$, and so $y \leq 13^{n^2+n+1}$.

The relevance of Theorem D to the problem at hand is revealed in the exercises that follow.

Exercise 1. Show that for any function $f(x_1, \ldots, x_n)$, if the relation $f(x_1, \ldots, x_n) = y$ is Σ_1, then so is the relation $f(x_1, \ldots, x_n) \neq y$ (and, thus, any Σ_1-function is recursive). [Hint: If $f(x_1, \ldots, x_n)$ is unequal to y, then it is equal to some number other than y.]

Exercise 2. Prove that the function $\pi(x)$ is recursive.

Exercise 3. Show that for any recursive relation $R(x, y)$ and any recursive function $f(x)$, the relation $(\exists y \leq f(x))R(x, y)$ is recursive.

Exercise 4. The condition Seqt(x) (x is the Gödel number of a formation sequence for terms) is obviously recursive (it is even Σ_0). The condition tm(x) (E_x is a term) has been shown to be Σ_1. [We recall that tm$(x) \leftrightarrow \exists y(\text{Seqt}(y) \wedge x \in y)$.] Now show that the condition tm(x) is recursive by showing that

$$\text{tm}(x) \leftrightarrow (\exists y \leq \pi(x))(\text{Seqt}(y) \wedge x \in y).$$

[Hint: The definition of Seqt(x) involves the relation $R_1(x, y, z)$ (mirroring the formation relation for terms) which has the property that if $R_1(x, y, z)$ holds, then z must be greater than both x and y. Therefore, E_x is a term iff it is a member of a formation sequence $(E_{a_1}, \ldots, E_{a_k})$ *without repetitions* and such that each a_i is $\leq x$ (and hence $k \leq x$, since there are no repetitions), and so by Theorem A, the sequence number of $(a_1, \ldots, a_k)$ is $\leq x$.]

Exercise 5. Similarly, show that in item (11) of Prop. 3, Ch. 3, if we replace "$\exists y$" by "$(\exists y \leq \pi(x))$", the equivalence still holds and, therefore, the set of Gödel numbers of formulas is recursive.

Exercise 6. Now show that all the items (1)–(15) of Prop. 3, Ch. 3 are recursive.

Exercise 7. Consider now the relation Fr(x, y) (cf. Exercise 1 of Ch. 3). Using the function $\pi(x)$, show that this relation is recursive. Then show that the set of Gödel numbers of sentences is recursive.

Exercise 8.
(1) By the (universal) closure of a formula

$$F(v_{i_1}, \ldots, v_{i_n}),$$

we mean the sentence $\forall v_{i_1} \forall v_{i_2} \ldots \forall v_{i_n} F(v_{i_1}, v_{i_2}, \ldots, v_{i_n})$. [If F has no free variables, then F is its own closure.] Show that the following relation between x and y is recursive. E_x is a formula and E_y is its closure.

(2) Using (1), show that for any system S, the set of Gödel numbers of the provable formulas is r.e. iff the set of Gödel numbers of the provable sentences is r.e., and the one set is recursive iff the other set is recursive.

Chapter V

Gödel's Proof Based on ω-Consistency

The proof that we have just given of the incompleteness of Peano Arithmetic was based on the underlying assumption that Peano Arithmetic is *correct*—i.e., that every sentence provable in P.A. is a *true* sentence. Gödel's original incompleteness proof involved a much weaker assumption—that of ω-consistency to which we now turn.

We consider an arbitrary axiom system S whose formulas are those of Peano Arithmetic, whose axioms include all those of Groups I and II (or alternatively, any set of axioms for first-order logic with identity such that all logically valid formulas are provable from them), and whose inference rules are modus ponens and generalization. (It is also possible to axiomatize first-order logic in such a way that modus ponens is the only inference rule—cf. Quine [1940].) In place of the axioms of Groups III and IV, however, we can take a completely arbitrary set of axioms. Such a system S is an example of what is termed a *first-order theory*, and we will consider several such theories other than Peano Arithmetic. (For the more general notion of a first-order theory, the key difference is that we do not necessarily start with $+$ and $\times$ as the undefined function symbols, nor do we necessarily take $\leq$ as the undefined predicate symbol. Arbitrary function symbols and predicate symbols can be taken, however, as the undefined function and predicate symbols—cf. Tarski [1953] for details. However, the only theories (or "systems", as we will call them) that we will have occasion to consider are those whose formulas are those of P.A.)

S is called *simply consistent* (or just "consistent" for short) if no sentence is both provable and refutable in S. Now, S is called ω-*inconsistent* if there is a formula $F(w)$ in one free variable w, such that the sentence $\exists w F(w)$ is provable, yet all the sentences

$F(\bar{0}), F(\bar{1}), \ldots, F(\bar{n}), \ldots$ are refutable. Of course, such a system cannot be correct because if $\exists w F(w)$ is true, then for at least one n, the sentence $F(\bar{n})$ must be true, and so a correct system cannot be ω-inconsistent. It is possible, however, for an ω-inconsistent system to be simply consistent, as we will see. A system S is called ω-*consistent* if it is not ω-inconsistent—in other words, if whenever a sentence $\exists w F(w)$ is provable in S, then for at least one number n, the sentence $F(\bar{n})$ is not refutable in S.

If S is (simply) inconsistent, then every sentence is provable in S (since S contains all axioms and inference rules of propositional logic), and hence it must be ω-inconsistent. Stated otherwise, if S is ω-consistent, then it is also simply consistent.

A system S is called *recursively axiomatizable* or just "axiomatizable" for short if the set P of Gödel numbers of the provable formulas of S is Σ_1. Axiomatizable systems are also called *formal systems* or *r.e. systems* ("r.e." for recursively enumerable). They may also be called Σ_1-*systems*. We proved in the last chapter that the system P.A. is axiomatizable. An example of a system that is not axiomatizable is the system $\mathcal{N}$ (the complete theory of arithmetic) whose non-logical axioms are all *correct* arithmetic formulas. (The provable formulas of $\mathcal{N}$ are nothing more than the axioms of $\mathcal{N}$ since any logical consequence of correct formulas is again a correct formula.) The set of Gödel numbers of the provable formulas of $\mathcal{N}$ is not even arithmetic, let alone Σ_1, so the system $\mathcal{N}$ is certainly not axiomatizable.

Given two systems S and S_1, we call S_1 a *subsystem* of S or S an *extension* of S_1 if all provable formulas of S_1 are also provable in S. As an example, the system P.A. is a subsystem of the complete system $\mathcal{N}$ (since P.A. is correct). We will later on consider in turn some significant subsystems of P.A.

As we have remarked, every correct system is automatically ω-consistent. Hence the assumption that the system P.A. is correct is stronger than the assumption that P.A. is ω-consistent. We have proved the incompleteness of P.A. under the assumption that P.A. is correct. A major purpose of this chapter is to prove the following:

Theorem G.[1]. *If Peano Arithmetic is ω-consistent, then it is incomplete.*

We will establish Theorem G as a consequence of the following two theorems.

---

[1] Gödel's original version of the Incompleteness Theorem for Peano Arithmetic

Theorem A.[2] *If S is any axiomatizable ω-consistent system in which all true Σ_0-sentences are provable, then S must be incomplete.*

Theorem B. *All true Σ_0-sentences are provable in* P.A.

Since we have already proved that P.A. is axiomatizable, Theorem G will follow as soon as we have proved Theorem A and Theorem B.

Discussion. The proof of Theorem G amounts to far more than just "another proof" that P.A. is incomplete. Out of the proof will emerge much valuable information about P.A. and some of its subsystems which will be crucial for the rest of this volume.

In part I of this chapter, we will derive Theorem A as a consequence of some still more abstract incompleteness theorems—theorems that apply to non-axiomatizable systems as well as to axiomatizable systems. Part II will be devoted to the proof of Theorem B—in fact to the proof of some significantly stronger results.

I. *Some Abstract Incompleteness Theorems*

§1. A Basic Incompleteness Theorem.

A formula $F(v_1)$ (with v_1 as the only free variable) will be said to *represent* a number set A in S if A consists of those and only those numbers n such that the sentence $F(\overline{n})$ is *provable* in S. More generally, a formula $F(v_1, \ldots, v_k)$ *represents* the set of all k-tuples $(n_1, \ldots, n_k)$ such that $F(\overline{n}_1, \ldots, \overline{n}_k)$ is provable in S.

Consider now the case where S is the system P.A. A formula $F(v_1)$ *expresses* the set of all n such that $F(\overline{n})$ is a *true* sentence, whereas $F(v_1)$ *represents* (in P.A.), the set of all n such that $F(\overline{n})$ is *provable* in P.A. Since P.A. is correct, the set represented by $F(v_1)$ is a subset of the set expressed by $F(v_1)$, but the difference between these two sets can be quite drastic. For example, let G be a true sentence which is not provable in P.A., and let $F(v_1)$ be the formula $G \wedge (v_1 = v_1)$. The set expressed by $F(v_1)$ is the set of all natural numbers, but the set *represented* by $F(v_1)$ in P.A. is only the empty set! (Why?) In general, the expressible sets are the arithmetic sets, whereas the representable sets of P.A. will turn out to be only the Σ_1-sets (as we

---

[2] A Generalization of Gödel's Theorem

I. Some Abstract Incompleteness Theorems

will see). We note that the expressible sets are those representable in the complete theory $\mathcal{N}$ (and, thus, expressibility is but a special case of representablity — a fact noted by Mostowski [1961]).

We now consider an arbitrary system $\mathcal{S}$ (not necessarily axiomatizable). We let P be the set of Gödel numbers of the provable formulas of $\mathcal{S}$ and R be the set of Gödel numbers of the refutable formulas of $\mathcal{S}$. As with the special case of P.A., we let P^* be the set of all n such that $E_n[\overline{n}]$ is provable in $\mathcal{S}$, and we let R^* be the set of all n such that $E_n[\overline{n}]$ is refutable in $\mathcal{S}$. As with the special case of P.A., the sets P^* and R^* will play the principal roles in the drama about to unfold.

Theorem 1. *Suppose $\mathcal{S}$ is simply consistent and $H(v_1)$ is a formula whose negation represents the set P^* in $\mathcal{S}$. Then the sentence $H(\overline{h})$ is neither provable nor refutable in $\mathcal{S}$ where h is the Gödel number of the formula $H(v_1)$.*

Before proving Theorem 1, we shall explicitly note some basic facts. For any formula $H(v_1)$ and any number n, the sentence $H(\overline{n}) \equiv H[\overline{n}]$ (unabbreviated: $H(\overline{n}) \equiv \forall v_1(v_1 = \overline{n} \supset H(v_1))$) is not only arithmetically true, but is a theorem of first-order logic with identity. Hence it is actually provable in $\mathcal{S}$ (since all axioms of Groups I and II are axioms of $\mathcal{S}$). Therefore, $H(\overline{n})$ is provable in $\mathcal{S}$ iff $H[\overline{n}]$ is provable in $\mathcal{S}$, and $H(\overline{n})$ is refutable in $\mathcal{S}$ iff $H[\overline{n}]$ is refutable in $\mathcal{S}$. In particular, if h is the Gödel number of $H(v_1)$, then $H(\overline{h})$ is provable in $\mathcal{S}$ iff $H[\overline{h}]$ is provable in $\mathcal{S}$ iff $h \in P^*$. Similarly, $H(\overline{h})$ is refutable in $\mathcal{S}$ iff $h \in R^*$. And so we have:

Lemma 1. *For any formula $H(v_1)$ with Gödel number h:*
1. *$H(\overline{h})$ is provable in $\mathcal{S} \leftrightarrow h \in P^*$.*
2. *$H(\overline{h})$ is refutable in $\mathcal{S} \leftrightarrow h \in R^*$.*

Proof of Theorem 1. Assume hypothesis. Since the *negation* of $H(v_1)$ represents P^*, then for any number n, the sentence $H(\overline{n})$ is *refutable* in $\mathcal{S}$ iff $n \in P^*$. In particular, $H(\overline{h})$ is refutable in $\mathcal{S}$ iff $h \in P^*$. But it is also the case that $H(\overline{h})$ is *provable* in $\mathcal{S}$ iff $h \in P^*$ (by Lemma 1). Therefore, $H(\overline{h})$ is provable in $\mathcal{S}$ iff $H(\overline{h})$ is refutable in $\mathcal{S}$. This means that $H(\overline{h})$ is either both provable and refutable in $\mathcal{S}$ or neither. By the assumption of simple consistency, $H(\overline{h})$ is not both provable and refutable in $\mathcal{S}$; hence it is neither.

Question. Instead of taking a formula whose negation represents P^*, what about taking a formula that represents $\widetilde{P^*}$? This strategy is quite useless! Why? [The answer will emerge from Exercise 2

Corollary. *If P^* is representable in S and S is consistent, then S is incomplete.*

Proof 1. Suppose S is consistent and $F(v_1)$ is a formula that represents P^*. Then the formula $\sim\sim F(v_1)$ also represents P^* (why?). Hence the formula $H(v_1)$—viz. $\sim F(v_1)$—is a formula whose negation represents P^*, and so the hypothesis of Theorem 1 is fulfilled.

Proof 2. The proof above, though, following from Theorem 1 and avoiding a separate diagonal argument, is unnecessarily roundabout (and also is not applicable to systems based on intuitionistic logic in which we don't have the double negation principle). The following argument is more direct.

Suppose $H(v_1)$ represents P^*. Let k be the Gödel number of the formula $\sim H(v_1)$. Then (assuming S is consistent) the sentence $H(\overline{k})$ is undecidable in S. We leave the proof of this to the reader.

Exercise 1. Prove that the sentence $H(\overline{k})$ above is undecidable in S (if S is consistent).

Exercise 2. Prove that it is impossible that the set $\widetilde{P^*}$ is representable in S (regardless of whether S is consistent or not). (This answers the question following the proof of Theorem 1.)

Exercise 3. Although $\widetilde{P}^*$ is not representable in S, it might happen that it is representable in some consistent extension S' of S. Show that if it is, then S must be incomplete. How does this relate to any theorems previously proved?

A Dual of Theorem 1. In T.F.S. we proved the following theorem—a "dual" of Theorem 1.

Theorem 1°. *If the set R^* is representable in S and S is simply consistent, then S is incomplete.*

Proof. Suppose S is consistent and $H(v_1)$ represents R^* in S. Again let h be the Gödel number of $H(v_1)$. Then $H(\overline{h})$ is provable $\leftrightarrow h \in R^* \leftrightarrow H(\overline{h})$ is refutable (by Lemma 1). The rest of the argument is the same as that of Theorem 1.

Theorem 1 (or alternatively Thm. 1°) provides a path to the proof of Theorem A (and hence to Theorem G). In the next section we will show that if the hypothesis of Theorem A holds, then the sets P^* and R^* are both representable in S.

I. Some Abstract Incompleteness Theorems

Remark. Theorems 1° and 1 are respectively special cases of Exercises 2 and 4 of Chapter I. Can the reader see why?

Exercise 4. Suppose S is a subsystem of $\mathcal{N}$ (i.e., all sentences provable in S are true). Let $H(v_1)$ be a formula in v_1 and h be its Gödel number.

1. Suppose the negation of $H(v_1)$ represents and expresses the set P^*. We know that the sentence $H(\overline{h})$ is undecidable in S (S is automatically consistent since it is a subsystem of $\mathcal{N}$). Nevertheless, $H(\overline{h})$ is either true or false. Which is it?
2. Suppose $H(v_1)$ represents and expresses R^*. Is the sentence $H(\overline{h})$ true or false?

§2. The ω-Consistency Lemma.

Now we will see how ω-consistency enters the picture.

We shall say that a formula $F(v_1, v_2)$ *enumerates* a set A in S if for every number n, the following two conditions hold:

1. If $n \in A$, then there is at least one number m such that the sentence $F(\overline{n}, \overline{m})$ is provable in S.
2. If $n \notin A$, then for *every* number m, the sentence $F(\overline{n}, \overline{m})$ is refutable in S.

We say that A is enumerable in S iff there is some formula $F(v_1, v_2)$ that enumerates A in S.

Lemma ω—The ω-consistency lemma. *If S is ω-consistent, then every set enumerable in S is representable in S. More specifically, suppose S is ω-consistent and that $F(v_1, v_2)$ is a formula that enumerates A in S. Then the formula $\exists v_2 F(v_1, v_2)$ represents A in S.*

Proof. Assume hypothesis.

1. Suppose $n \in A$. Then for some m, the sentence $F(\overline{n}, \overline{m})$ is provable in S. Hence the sentence $\exists v_2 F(\overline{n}, v_2)$ is provable in S.
2. Conversely, suppose $\exists v_2 F(\overline{n}, v_2)$ is provable in S. If n is not in A, then all the sentences $F(\overline{n}, \overline{0}), F(\overline{n}, \overline{1}), \ldots, F(\overline{n}, \overline{m}), \ldots$, would be refutable in S, which would mean that S is ω-inconsistent (since $\exists v_2 F(\overline{n}, v_2)$ is provable in S). So if S is ω-consistent, then n must be in A.

By (1) and (2), the formula $\exists v_2 F(v_1, v_2)$ represents A in S.

From Theorem 1, Cor., and Theorem 1° and the above lemma, we have:

Theorem 2. *If either P^* or R^* is enumerable in S and S is ω-consistent, then S is incomplete.*

Exercise 5. Prove that if all true Σ_0-sentences are provable in S and S is ω-consistent, then all Σ_1-sets are representable in S.

Exercise 6. Suppose $F(v_1, v_2)$ is a formula that represents in S the same relation that it expresses and that for all numbers n and m, the sentence $F(\overline{n}, \overline{m})$ is either provable or refutable in S. Prove that if S is ω-consistent, then the formula $\exists v_2 F(v_1, v_2)$ represents in S the same set that it expresses.

A Sharpening of Theorem 2. Following Gödel, we now prove a significantly sharper form of Theorem 2. Suppose S is ω-consistent and that the set P^* is enumerable in S. Let $A(v_1, v_2)$ be a formula that enumerates P^* in S. Then the formula $\exists v_2 A(v_1, v_2)$ represents P^* in S (by the ω-consistency lemma). The formula $\exists v_2 A(v_1, v_2)$ in unabbreviated notation is $\sim \forall v_2 \sim A(v_1, v_2)$. Therefore, the formula $\forall v_2 \sim A(v_1, v_2)$ is a formula whose *negation* represents P^* in S. So by Theorem 1, the sentence $\forall v_2 \sim A(\overline{a}, v_2)$ is undecidable in S where a is the Gödel number of the formula $\forall v_2 \sim A(v_1, v_2)$.

Now comes a highly significant point: Let G be the sentence

$$\forall v_2 \sim A(\overline{a}, v_2)$$

(this G is Gödel's sentence!). We have just seen that if S is ω-consistent, then G is neither provable nor refutable in S. However, only the *simple* consistency of S is required to show that G is not provable in S! Here is the reason why.

Suppose the sentence $\forall v_2 \sim A(\overline{a}, v_2)$ is provable in S. Then a is in P^* (by Lemma 1, taking $\forall v_2 A(v_1, v_2)$ for $H(v_1)$). Since $A(v_1, v_2)$ enumerates the set P^* in S, there must be a number m such that the sentence $A(\overline{a}, \overline{m})$ is provable in S. Hence the sentence $\exists v_2 A(\overline{a}, v_2)$ is provable in S, but this sentence is $\sim \forall v_2 \sim A(\overline{a}, v_2)$, which is the negation of $\forall v_2 \sim A(\overline{a}, v_2)$—i.e., it is the sentence $\sim G$. So if G is provable in S, so is its negation, which means that S is then *simply* inconsistent. So if S is simply consistent, then G is not provable in S. [The assumption of ω-consistency is required only to show that G is not refutable in S.] We have, thus, proved:

Theorem 3. *Suppose $A(v_1, v_2)$ enumerates P^* in S. Let a be the Gödel number of the formula $\forall v_2 \sim A(v_1, v_2)$ and let G be the sen-*

I. Some Abstract Incompleteness Theorems

tence $\forall v_2 \sim A(\bar{a}, v_2)$. Then:

1. If S is simply consistent, then G is not provable in S.
2. If S is ω-consistent, then it is also the case that G is not refutable in S.

Of course, there is also a "dual" result for the set R^* whose proof we leave to the reader:

Theorem 3°. Suppose $B(v_1, v_2)$ is a formula that enumerates R^* in S, b is the Gödel number of the formula $\exists v_2 B(v_1, v_2)$ and S is the sentence $\forall v_2 \sim B(\bar{b}, v_2)$. Then:

1. If S is simply consistent, then S is not provable in S.
2. If S is ω-consistent, then S is neither provable nor refutable.

Theorem A (which we are aiming at) will follow easily as a consequence of the following theorem, which is now within reach.

Theorem A'. If S is any axiomatizable ω-consistent system in which all Σ_1-sets are enumerable, then S must be incomplete.

Proof. Assume hypothesis. Since S is axiomatizable, the set P is Σ_1 (by definition of axiomatizable). Therefore, the set P^* is Σ_1 (we proved in the last chapter that for any Σ_1-set A, the set A^* is also Σ_1). Then by hypothesis, the set P^* is enumerable in S. The conclusion then follows by Theorem 3 (also by Theorem 3°).

Remark. If the set P is Σ_1, then so is the set R (why?) and, therefore, so is the set R^*. Hence the dual argument of Theorem 3° can also be used to prove Theorem A'.

To establish Theorem A as a consequence of Theorem A', all that remains is to show that if all true Σ_0-sentences are provable in S, then all Σ_1-sets are enumerable in S. As we are at it, we will show something more general. Let us say that a formula $F(v_1, \ldots, v_n, v_{n+1})$ *enumerates* a relation $R(x_1, \ldots, x_n)$ in S if for all numbers $k_1, \ldots, k_n$ the following two conditions hold:

1. If $R(k_1, \ldots, k_n)$ holds, then there is a number k such that the sentence $F(\bar{k}_1, \ldots, \bar{k}_n, \bar{k})$ is provable in S;
2. If $R(k_1, \ldots, k_n)$ doesn't hold, then for every k, the sentence $F(\bar{k}_1, \ldots, \bar{k}_n, \bar{k})$ is refutable in S.

Lemma 2. If all true Σ_0-sentences are provable in S, then all Σ_1-sets and relations are enumerable in S.

Proof. Assume hypothesis. Let $R(x_1,\ldots,x_n)$ be any Σ_1-relation (or set, if $n=1$). Then there is a Σ_0-relation $S(x_1,\ldots,x_n,y)$ such that for all $x_1,\ldots,x_n$,

$$R(x_1,\ldots,x_n) \leftrightarrow \exists y S(x_1,\ldots x_n, y).$$

Let $F(v_1,\ldots v_n, v_{n+1})$ be a Σ_0-formula that expresses the relation $S(x_1,\ldots,x_n,y)$. We show that $F(v_1,\ldots,v_n,v_{n+1})$ enumerates the relation $R(x_1,\ldots,x_n)$ in $\mathcal{S}$.

1. Suppose $R(k_1,\ldots,k_n)$ holds. Then for some number k,

$$S(k_1,\ldots,k_n,k)$$

holds. Hence $F(\overline{k}_1,\ldots,\overline{k}_n,\overline{k})$ is a true Σ_0- sentence, it is provable in $\mathcal{S}$ (by hypothesis).

2. Suppose $R(k_1,\ldots,k_n)$ doesn't hold. Then for every number k, it is false that $S(k_1,\ldots,k_n,k)$. Hence for every k, the sentence $F(\overline{k}_1,\ldots,\overline{k}_n,\overline{k})$ is false, so the sentence

$$\sim F(\overline{k}_1,\ldots,\overline{k}_n,\overline{k})$$

is true, and being a Σ_0-sentence, it is provable in $\mathcal{S}$, which means that $F(\overline{k}_1,\ldots,\overline{k}_2,\overline{k})$ is refutable in $\mathcal{S}$ for every k.

By (1) and (2), the formula $F(v_1,\ldots,v_n,v_{n+1})$ enumerates

$$R(x_1,\ldots,x_n)$$

in $\mathcal{S}$.

Having proved the above lemma and Theorem A′, our proof of Theorem A is complete.

Before we turn to the proof of Theorem B, we wish to pause and observe the following curious "self-strengthening" of Theorem A.

Theorem A*. *If $\mathcal{S}$ is any axiomatizable ω-consistent system in which no false Σ_0-sentence is provable, then $\mathcal{S}$ is incomplete.*

Proof. Assume hypothesis. Now, it either is or isn't the case that all true Σ_0-sentences are provable in $\mathcal{S}$. If it is the case, then $\mathcal{S}$ is incomplete by Theorem A. If it isn't the case, then some true Σ_0-sentence X is not provable in $\mathcal{S}$. Then the sentence $\sim X$ is a false Σ_0-sentence and is not provable in $\mathcal{S}$ (by hypothesis). Hence X is undecidable in $\mathcal{S}$.

It is curious that we used Theorem A to prove Theorem A* whose hypothesis is weaker (weaker, because if $\mathcal{S}$ is ω-consistent, then it is also simply consistent. Hence if all true Σ_0-sentences are provable in $\mathcal{S}$, then for any false Σ_0-sentence X, it cannot be provable in $\mathcal{S}$ since its negation is a true Σ_0-sentence provable in $\mathcal{S}$). As indicated in Exercise 7 below, Theorem A* can be proved directly (moreover in a manner that gives more information than the above proof), and Theorem A can then be obtained as a corollary.

Exercise 7. Suppose $\mathcal{S}$ is axiomatizable and $A(v_1, v_2)$ is a Σ_0-formula that expresses a Σ_0-relation $R(x, y)$ whose domain is P^*. Let a be the Gödel number of the formula $\forall v_2 \sim A(v_1, v_2)$, and let G be the sentence $\forall v_2 \sim A(\bar{a}, v_2)$.

1. Show that if G is provable in $\mathcal{S}$, then some false Σ_0-sentence is provable in $\mathcal{S}$.
2. Show that if G is refutable in $\mathcal{S}$ and $\mathcal{S}$ is ω-consistent, then some true Σ_0-sentence fails to be provable in $\mathcal{S}$.
3. Show that Theorem A* follows from (1) and (2).

Solution of Exercise 7. This result does not appear to be generally known, and so we will give the solution.

1. Suppose G is provable in S. Then $a \in P^*$. Hence for some n, the sentence $A(\bar{a}, \bar{n})$ is true, and for such an n, the sentence $\sim A(\bar{a}, \bar{n})$ is false. But for every n, the sentence $\sim A(\bar{a}, \bar{n})$ is provable (since $\forall y \sim A(\bar{a}, y)$ is provable). Hence for at least one number n, the false Σ_0-sentence $\sim A(\bar{a}, \bar{n})$ is provable.
2. Suppose $\sim G$ is provable—i.e. $\sim \forall y \sim A(\bar{a}, y)$ is provable. Thus, $\exists y A(\bar{a}, y)$ is provable. Now suppose S is ω-consistent. Then for at least one number n, the sentence $\sim A(\bar{a}, \bar{n})$ is *not* provable. But since S is ω-consistent, it is also simply consistent, and so G is not provable. Therefore, $a \notin P^*$. So for every n, the sentence $A(\bar{a}, \bar{n})$ is false, and for every n, the sentence $\sim A(\bar{a}, \bar{n})$ is true. Yet for some n, the sentence $\sim A(\bar{a}, \bar{n})$ is not provable, and so at least one true Σ_0-sentence fails to be provable.
3. Suppose now all true Σ_0-sentences are provable in S. If S is simply consistent, then no false Σ_0-sentence is provable in S. Hence by (1), the sentence G is not provable. Since no true Σ_0-sentence fails to be provable in S, it follows from (2) that if G is refutable in S, then S must be ω-inconsistent.

Since we know that P.A. is axiomatizable, it follows by Theorem A* that if P.A. is complete, then either it is ω-inconsistent (in fact with respect to some Σ_1-formula) or some false Σ_0-sentence is provable in P.A. After we have proved Theorem B, we will know that if P.A. is complete, then it is ω-inconsistent.

II. Σ_0-Completeness

We now turn to the proof of Theorem B—in fact to the proof of some stronger results that will be needed in several subsequent chapters.

We will call a system S Σ_0-*complete* if all true Σ_0-sentences are provable in S. We now wish to prove the Σ_0-completeness not only of P.A. but also of several subsystems of P.A. which play an important role in metamathematics.

§3. First we turn to a useful sufficient condition that a system S be Σ_0-complete. We shall say that a Σ_0-sentence is *correctly decidable* in S if it is either true and provable in S or false and refutable in S.

Proposition 1. *The following two conditions are jointly sufficient for S to be Σ_0-complete.*

C_1: *Every atomic Σ_0-sentence is correctly decidable in S.*
C_2: *For any Σ_0-formula $F(w)$ with w as the only free variable and for every number n, if the sentences $F(\overline{0}), \ldots, F(\overline{n})$ are all provable in S, then so is the sentence $(\forall w \leq \overline{n})F(w)$.*

Proof. Suppose C_1 and C_2 both hold. We show by induction on degrees of sentences that all Σ_0-sentences are correctly decidable in S (which, of course, implies that all true Σ_0-sentences are provable in S.)

1. By C_1, all Σ_0-sentences of degree 0 are correctly decidable in S.
2. It is obvious by propositional logic that for any sentences X and Y, if X and Y are both correctly decidable in S, then $\sim X$ and $X \supset Y$ are correctly decidable in S. We leave the proof of this to the reader.
3. Any Σ_0-sentence S, which is neither atomic nor of the form $\sim X$ or $X \supset Y$, must be of the form $(\forall w \leq \overline{n})F(w)$ where $F(w)$ is a Σ_0-formula of lower degree than S and contains w as the only free variable. So we consider now a Σ_0-sentence $(\forall w \leq \overline{n})F(w)$ such that all Σ_0-sentences of lower degree are correctly decidable in S. We must show that $(\forall w \leq \overline{n})F(w)$ is

II. Σ_0-Completeness

also correctly decidable in S.

Suppose the sentence is true. Then each of the sentences

$$F(\overline{0}), \ldots, F(\overline{n})$$

is true. Hence each of them is provable in S (by induction hypothesis) since they are of lower degree than $(\forall w \leq \overline{n})F(w)$. Then by condition C_2, the sentence $(\forall w \leq \overline{n})F(w)$ is provable in S.

Suppose the sentence is false. Then for at least one $m \leq n$, the sentence $F(\overline{m})$ is false and, hence, refutable in S (again by induction hypothesis). Since $\overline{m} \leq \overline{n}$ is a true Σ_0-sentence, it is provable in S (by C_1). Since $\overline{m} \leq \overline{n}$ and $\sim F(\overline{m})$ are both provable, $\overline{m} \leq \overline{n} \supset F(\overline{m})$ is refutable in S. But

$$(\forall w \leq \overline{n})F(w) \supset (\overline{m} \leq \overline{n} \supset F(\overline{m}))$$

is provable (it is logically valid), and so $(\forall w \leq \overline{n})F(w)$ is refutable in S. This completes the proof.

Proposition 2. *The following three conditions are jointly sufficient for S to be Σ_0-complete.*

D_1: *All true atomic Σ_0-sentences are provable in S.*

D_2: *For any distinct numbers m and n, the sentence $\overline{m} \neq \overline{n}$ is provable in S.*

D_3: *For any variable w and any number n, the formula*

$$w \leq \overline{n} \supset (w = \overline{0} \vee \ldots \vee w = \overline{n})$$

is provable in S.

More specifically, conditions D_1, D_2 and D_3 jointly imply condition C_1 of Proposition 1, and D_3 implies condition C_2.

Proof.

1. Suppose conditions D_1, D_2 and D_3 hold. We show that condition C_1 must hold.

   By D_1, all true atomic Σ_0-sentences are provable in S. It remains to show that all false atomic Σ_0-sentences are refutable in S. If the false sentence is of the form $\overline{m} = \overline{n}$, then it is refutable in S by D_2. Suppose the false sentence is of the form $\overline{m} \leq \overline{n}$. Since it is false, then all the sentences $\overline{m} = \overline{0}, \ldots, \overline{m} = \overline{n}$ are false. Hence they are all refutable in S (by D_2), and the sentence $\overline{m} = \overline{0} \vee \ldots \vee \overline{m} = \overline{n}$ is refutable. It then follows from D_3 (substituting $\overline{m}$ for w) that $\overline{m} \leq \overline{n}$ is refutable.

Consider now a false Σ_0-sentence of the form $\overline{m} + \overline{n} = \overline{k}$. Since the sentence is false, $m + n = p$ for some number $p \neq k$. Then $\overline{m} + \overline{n} = \overline{p}$ is provable in S (by D_1), and $\overline{p} \neq \overline{k}$ is provable in S (by D_2). Therefore, the formula $\overline{m} + \overline{n} \neq \overline{k}$ is provable in S (because for any terms t_1, t_2 and t_3, the formula

$$(t_1 = t_2 \wedge t_2 \neq t_3) \supset (t_1 \neq t_3)$$

is a valid formula of first order logic with identity and is, hence, provable in S). Thus, the false sentence $\overline{m} + \overline{n} = \overline{k}$ is refutable in S. The proof that every false Σ_0-sentence of the form $\overline{m} \cdot \overline{n} = \overline{k}$ is refutable in S is similar.

2. We now show that condition C_2 is derivable from D_3. Assume D_3. Suppose $F(w)$ is a Σ_0-formula (with w as the only free variable), and n is a number such that each of the n sentences $F(\overline{0}), \ldots, F(\overline{n})$ are all provable in S. Then the open formulas

$$w = \overline{0} \supset F(w), \ldots, w = \overline{n} \supset F(w)$$

(being respective logical consequences of $F(\overline{0}), \ldots, F(\overline{n})$) are all provable in S. Therefore by propositional logic, the formula

$$w = 0 \vee \ldots \vee w = \overline{n} \supset F(w)$$

is provable in S and it follows (by propositional logic) that the formula $w \leq \overline{n} \supset F(w)$ is provable in S. Hence by the generalization rule, the sentence $\forall w(w \leq \overline{n} \supset F(w))$ is provable in S, and this sentence is the sentence $(\forall w \leq \overline{n})F(w)$.

§4. Some Σ_0-complete Subsystems of Peano Arithmetic.

We let (Q) be the system P.A. with axiom scheme N_{12} (the induction scheme) deleted. Thus, (Q) has only finitely many nonlogical axioms, to wit, the following nine:

N_1: $v_1' = v_2' \supset v_1 = v_2$.
N_2: $\sim v_1' = \overline{0}$.
N_3: $v_1 + \overline{0} = v_1$.
N_4: $v_1 + v_2' = (v_1 + v_2)'$.
N_5: $v_1 \cdot \overline{0} = \overline{0}$.
N_6: $v_1 \cdot v_2' = (v_1 \cdot v_2) + v_1$.
N_7: $v_1 \leq \overline{0} \equiv v_1 = \overline{0}$.
N_8: $v_1 \leq v_2' \equiv (v_1 \leq v_2 \vee v_1 = v_2')$.

II. Σ_0-Completeness

N_9: $v_1 \leq v_2 \vee v_2 \leq v_1$.

The system (Q) is a variant of one due to Raphael Robinson [1950] which plays an important role in modern research. We wish to show that all true Σ_0-sentences are not only provable in P.A., but even provable in (Q). As a matter of fact, axiom N_9 is not necessary for the proof of this result, and so we let (Q_0) be the system (Q) with axiom N_9 deleted. Thus, the nonlogical axioms of (Q_0) are the eight axioms N_1–N_8, and we will show that the system (Q_0) is Σ_0-complete.

We will show something still stronger. Another axiom system that has played an important role in the last three decades or so is the system (R) (also due to Raphael Robinson). This system has infinitely many non-logical axioms all subsumed under the following five axiom schemes.

Ω_1: All sentences $\overline{m} + \overline{n} = \overline{k}$, where $m + n = k$.
Ω_2: All sentences $\overline{m} \cdot \overline{n} = \overline{k}$, where $m \times n = k$.
Ω_3: All sentences $\overline{m} \neq \overline{n}$, where m and n are distinct numbers.
Ω_4: All formulas $v_1 \leq \overline{n} \equiv (v_1 = \overline{0} \vee \ldots \vee v_1 = \overline{n})$.
Ω_5: All formulas $v_1 \leq \overline{n} \vee \overline{n} \leq v_1$.

Finally, we let (R_0) be the system (R) with axiom scheme Ω_5 deleted. Our plan now is first to show that the system (R_0) is Σ_0-complete (which is almost immediate from Proposition 2) and then to show that (R_0) is a subsystem of (Q_0) (and also that (R) is a subsystem of (Q)).

Proposition 3. *The system (R_0) is Σ_0-complete.*

Proof. For any number n, the sentence $\overline{n} = \overline{n}$ is a theorem of first-order logic with identity. Hence it is provable in (R_0). Next, suppose $m \leq n$. Since $\overline{m} = \overline{m}$ is provable, so is

$$\overline{m} = \overline{0} \vee \ldots \vee \overline{m} = \overline{m} \vee \ldots \vee \overline{m} = \overline{n}.$$

Then using Ω_4, the sentence $\overline{m} \leq \overline{n}$ is provable. Thus, all true Σ_0 sentences of the form $\overline{m} \leq \overline{n}$ are provable in (R_0). Then by Ω_1 and Ω_2, it follows that all true atomic Σ_0-sentences are provable in (R_0). Hence condition D_1 of the hypothesis of Proposition 2 is satisfied for the system (R_0). Condition D_2 is also satisfied (by Ω_3). It remains to verify condition D_3. Since for each n, the formula

$$v_1 \leq \overline{n} \supset (v_1 = \overline{0} \vee \ldots \vee v_1 = \overline{n})$$

is provable in (R_0) (axiom scheme Ω_4), so is the sentence
$$\forall v_1(v_1 \leq \overline{n} \supset (v_1 = \overline{0} \vee \ldots \vee v_1 = \overline{n})).$$
Hence for any variable w, the formula $w \leq \overline{n} \supset (w = \overline{0} \vee \ldots \vee w = \overline{n})$ is provable in (R_0). The result then follows from Proposition 2.

Proposition 4. (R_0) *is a subsystem of* (Q_0).

Proof. Although the induction axioms of N_{12} are *not* axioms of (Q_0), we are perfectly free to use mathematical induction in our metalanguage to prove various things *about* the system (Q_0). The arguments that follow all use ordinary mathematical induction. The words "provable" and "refutable" shall now be understood to mean provable in (Q_0) and refutable in (Q_0).

1. Using axiom N_4, we can see that if $\overline{n} + \overline{m} = \overline{q}$ is provable in (Q_0), then so is $\overline{n} + \overline{m+1} = \overline{q+1}$. Also, $\overline{n} + \overline{0} = \overline{n}$ is provable (by N_3), and so we can successively prove
$$\overline{n} + \overline{1} = \overline{n+1}, \overline{n} + \overline{2} = \overline{n+2}, \ldots, \overline{n} + \overline{m} = \overline{n+m}.$$
[We have just used mathematical induction informally.] So all sentences of Ω_1 are provable in (Q_0).

2. Using axiom N_6, we see that if $\overline{n} \cdot \overline{m} = \overline{q}$ is provable, then so is
$$\overline{n} \cdot \overline{m+1} = \overline{q + \overline{n}},$$
and hence so is $\overline{n} \cdot \overline{m+1} = \overline{q+n}$ (since we have shown that $\overline{q} + \overline{n} = \overline{q+n}$ is provable). Thus, if $\overline{n} \cdot \overline{m} = \overline{n \times m}$ is provable, then so is $\overline{n} \cdot \overline{m+1} = \overline{n \times (m+1)}$. Also $\overline{n} \cdot \overline{0} = \overline{0}$ is provable (by N_5), and so we can successively prove
$$\overline{n} \cdot \overline{1} = \overline{n \times 1}, \overline{n} \cdot \overline{2} = \overline{n \times 2}, \ldots, \overline{n} \cdot \overline{m} = \overline{n \times m}.$$
Thus, all sentences of Ω_2 are provable in (Q_0).

3. For axiom scheme Ω_3, we are to show that for any number m and any *positive* number n, the sentence $\overline{m} \neq \overline{n+m}$ (and hence also the sentence $\overline{n+m} \neq \overline{m}$) is provable in (Q_0). We first note that for any numbers m and n, the sentence
$$\overline{m+1} = \overline{n+1} \supset \overline{m} = \overline{n}$$
is provable (by N_1). Hence
$$\overline{m} \neq \overline{n} \supset \overline{m+1} \neq \overline{n+1}$$
is provable. Therefore, if $\overline{m} \neq \overline{n}$ is provable, then so is
$$\overline{m+1} \neq \overline{n+1}.$$

Now, for any *positive* n, the sentence $\overline{0} \neq \overline{n}$ is provable (by N_2). Hence we can successively prove

$$\overline{1} \neq \overline{n+1}, \overline{2} \neq \overline{n+2}, \ldots, \overline{m} \neq \overline{n+m}.$$

Thus, all sentences of Ω_3 are provable in (Q_0).

4. For axiom scheme Ω_4, we show by induction on n that

$$v_1 \leq \overline{n} \equiv (v_1 = \overline{0} \vee \ldots \vee v_1 = \overline{n})$$

is provable in (Q_0). Well, $v_1 \leq \overline{0} \equiv v_1 = \overline{0}$ is provable (by N_7). Now suppose n is such that

$$v_1 \leq \overline{n} \equiv (v_1 = \overline{0} \vee \ldots \vee v_1 = \overline{n})$$

is provable. Then, since

$$v_1 \leq \overline{n+1} \equiv (v_1 \leq \overline{n} \vee v_1 = \overline{n+1})$$

is provable (by N_8), it follows by propositional logic that

$$v_1 \leq \overline{n+1} \equiv (v_1 \leq \overline{0} \vee \ldots \vee v_1 = \overline{n} \vee v_1 = \overline{n+1})$$

is provable. This completes the induction.

Proposition 5. (R) *is a subsystem of* (Q).

Proof. Since (R_0) is a subsystem of (Q_0), it is, of course, a subsystem of (Q). Also all formulas of Ω_5 are provable in (Q) (by N_9, substituting $\overline{n}$ for v_2) and so (R) is a subsystem of (Q).

We have now shown that (R_0) is Σ_0-complete and that (R_0) is a subsystem of (Q_0), which in turn is a subsystem of (Q), which in turn is a subsystem of P.A. Also (R_0) is a subsystem of (R). We, thus, have proved the following stronger version of Theorem B.

Theorem B⁺. *The systems* $(R_0), (R), (Q_0), (Q)$ *and P.A. are all Σ_0-complete.*

Exercise 8. Let Ω_4' be the axiom scheme

$$v_1 \leq \overline{n} \supset (v_1 = \overline{0} \vee \ldots \vee v_1 = \overline{n}),$$

and let (R') be the system (R) with Ω_4 replaced by the weaker scheme Ω_4'. Show that everything provable in (R) is provable in (R').

§5. Having proved Theorems A and B, our proof of Gödel's Theorem for P.A. (Theorem G) is complete. Let us review the broad outlines of the proof and discuss some related topics.

We have shown that all true Σ_0-sentences are provable in P.A. and hence that every Σ_1-set is *enumerable* in P.A. — enumerated, in fact, by a Σ_0-formula $F(v_1, v_2)$. The system P.A. is axiomatizable. Hence the set P is Σ_1, and so the set P^* is Σ_1. Therefore, there is a Σ_0-formula $A(v_1, v_2)$ (which, in fact, can be found by the methods of the last two chapters) which *enumerates* P^* in P.A. Then by the ω-consistency lemma, if P.A. is ω-consistent, then the formula $\forall v_2 \sim A(v_1, v_2)$ is a formula whose negation *represents* P^* in P.A. We let a be the Gödel number of this formula, and we let G be the sentence $\forall v_2 \sim A(\bar{a}, v_2)$. Then by Theorem 1, the sentence G is neither provable nor refutable in P.A. (assuming that P.A. is ω-consistent). It, furthermore, follows by Theorem 3 that only the *simple* consistency of P.A. is required to show that G is not provable in P.A. So if P.A. is simply consistent, then the sentence G is not provable in P.A. If P.A. is ω-consistent (which, of course, it is since it is correct), then $\sim G$ is not provable in P.A. either.

The sentence G, incidentally, is the *true* one of the pair (G and $\sim G$) (assuming that P.A. is consistent). This is a consequence of the following propositions.

Proposition 6. *For any simply consistent system S in which all true Σ_0-sentences are provable, all provable Σ_0-sentences are true. In particular, if P.A. is simply consistent, then all Σ_0-sentences provable in P.A. are true.*

Proof. Suppose all true Σ_0-sentences are provable in S. Suppose some false Σ_0-sentence X were provable in S. Then $\sim X$ is a true Σ_0-sentence and provable in S (by assumption). Hence X and $\sim X$ are both provable in S, and S is inconsistent. If S is consistent and Σ_0-complete, all Σ_0-sentences provable in S are true.

Proposition 7. *Let S be a system in which all true Σ_0-sentences are provable and which contains a Σ_0-formula $A(v_1, v_2)$ that enumerates the set P^*. Let G be the sentence $\forall v_2 \sim A(\bar{a}, v_2)$ where a is the Gödel number of $\forall v_2 \sim A(v_1, v_2)$. Then if S is simply consistent, then the sentence G is true.*

Proof. Assume hypothesis. By Theorem 3, the sentence G is not provable in S. Therefore, a is not in the set P^* (by Lemma 1). Since $A(v_1, v_2)$ enumerates P^* in S and $a \notin P^*$, for every number n, the sentence $A(\bar{a}, \bar{n})$ is refutable—i.e., $\sim A(\bar{a}, \bar{n})$ is provable. Then by Proposition 6, for every n, the sentence $\sim A(\bar{a}, \bar{n})$ is true. Hence, the universal sentence $\forall v_2 \sim A(\bar{a}, v_2)$ is true—this is the sentence G.

Corollary. *If P.A. is simply consistent, then Gödel's sentence G is true.*

We now *know* that Gödel's sentence G for Peano Arithmetic is true (since we know that P.A. is correct and consistent); we have informally "demonstrated" the truth of G, but the demonstration is evidently not formalizable in P.A. (cf. discussion of Gödel's Second Theorem in Ch. 9).

Exercise 9. Assuming P.A. to be consistent, suppose we add the sentence $\sim G$ as a new axiom and call the resulting system P.A. $+\{\sim G\}$. Show that this system is consistent but not ω-consistent. [This provides an example of an ω-inconsistent system, which is nevertheless simply consistent.]

§6. The ω-Incompleteness of P.A. Our first proof of the incompleteness of Peano Arithmetic (which used the Tarski truth set) was obviously far simpler than the proof of this chapter based on ω-consistency. However, the latter proof (in addition to being formalizable in P.A., which is necessary for Gödel's second theorem) reveals another property of Peano Arithmetic that is even more startling than that of incompleteness.

The average mathematician, not a specialist in mathematical logic, has at least heard the *statement* of Gödel's incompleteness theorem, though he or she may not have gone through the proof. But the average general mathematician appears not even to have heard of the still more remarkable fact that there is a formula $F(w)$, with one free variable w, such that all sentences $F(\bar{0}), F(\bar{1}), \ldots, F(\bar{n}), \ldots$ are provable in P.A., yet the universal sentence $\forall w F(w)$ is not provable in P.A.! This is the condition called *ω-incompleteness* by Tarski.

To see that this is so, recall that we have shown (under the assumption of simple consistency) that Gödel's sentence $\forall v_2 \sim A(\bar{a}, v_2)$ is not provable in P.A. Hence $a \notin P^*$ and all the sentences $\sim A(\bar{a}, \bar{0}), \sim A(\bar{a}, \bar{1}), \ldots, \sim A(\bar{a}, \bar{n}), \ldots$ are provable, yet the universal sentence $\forall v_2 \sim A(\bar{n}, v_2)$ is not provable in P.A.!

Of course this argument goes through for every consistent axiomatizable system in which all true Σ_0-sentences are provable, and so we have:

Theorem C—The ω-Incompleteness Theorem. *If S is any simply consistent axiomatizable system in which all true Σ_0-sentences are provable, then S is ω-incomplete.*

Exercise 10. For any sentence X, let us define $P(\overline{X})$ to be $P(\overline{x})$, where x is the Gödel number of X.

Under the assumption that P.A. is correct (or even ω-consistent), the Σ_1-formula $P(v_1)$ expresses the set P of Gödel numbers of the provable formulas of P.A. Thus, for any sentence X, the sentence $P(\overline{X})$ is true iff X is provable (in P.A.). We know that for any Σ_0-sentence X, if X is true, then X is provable—hence the sentence $X \supset P(\overline{X})$ is a *true* sentence. Show that for every Σ_0-sentence X, the sentence $X \supset P(\overline{X})$ is provable in P.A.

Exercise 11. Show that every true Σ_1-sentence is provable in P.A. [This implies that for every Σ_1-sentence X, the sentence $X \supset P(\overline{X})$ is true.] It is also the case that for every Σ_1-sentence X, the sentence $X \supset P(\overline{X})$ is provable in P.A., but the proof of this is extremely intricate and goes beyond the scope of this volume. The interested reader can find a sketch of this proof in Boolos [1979], Ch. 2.

Exercise 12. Prove that it is not the case that for every sentence X, the sentence $X \supset P(\overline{X})$ is provable in P.A. (assuming P.A. is correct).

Chapter VI

Rosser Systems

Our first proof of the incompleteness of P.A. was based on the assumption that P.A. is correct. Gödel's proof of the last chapter was based on the metamathematically weaker assumption that P.A. is ω-consistent. Rosser [1936] subsequently showed that P.A. can be proved incomplete under the still weaker metamathematical assumption that P.A. is simply consistent! Now, Rosser did not show that the Gödel sentence G of the last chapter is undecidable on the weaker assumption of simple consistency. He constructed another sentence (a more elaborate one) which he showed undecidable on the basis of simple consistency.

Our first proof of the incompleteness of P.A. boils down to finding a formula that *expresses* the set $\tilde{P}^*$ (or alternatively one that expresses R^*). Gödel's proof, which we gave in the last chapter, boils down to *representing* one of the sets P^* and R^* in P.A. and the only way known in Gödel's time of doing this involved the assumption of ω-consistency. [This assumption was not needed to show that the sets P^* and R^* are *enumerable* in P.A.—it was in passing from the enumerability of these sets to their representability that ω-consistency stepped in.] Now, Rosser did not achieve incompleteness by representing either of the sets P^* and R^*, but rather by representing some superset of R^* disjoint from P^*—this can be done under the weaker assumption of simple consistency—and it also serves to establish incompleteness, as we will see.

The axiom schemes Ω_4 and Ω_5 of the system (R) will play a key role in this and the next chapter. We shall say that a system S is an extension of Ω_4 and Ω_5 if all formulas of Ω_4 and Ω_5 are provable in S. We will prove the following theorem and its corollaries.

Theorem R. *Every simply consistent axiomatizable extension of Ω_4 and Ω_5 in which all Σ_1-sets are enumerable must be incomplete.*

Corollary 1. *Every simply consistent axiomatizable extension of Ω_4 and Ω_5 in which all true Σ_0-sentences are provable must be incomplete.*

Corollary 2. *Every simply consistent axiomatizable extension of the system (R) is incomplete.*

Corollary 3. *The system P.A., if simply consistent, is incomplete.*

Remark. Theorem R differs from Theorem A of the last chapter in that the ω-consistency hypothesis of Theorem A is weakened to simple consistency. To compensate, the additional assumption of $\mathcal{S}$ being an extension of Ω_4 and Ω_5 is required. The two theorems are apparently of incomparable strength. Similar remarks apply to Corollary 1 and Theorem A of the last chapter. Corollary 2 is referred to in Shoenfield [1967] as the *Gödel-Rosser Incompleteness Theorem*.

§1. Some Abstract Incompleteness Theorems After Rosser.

We showed in the last chapter (Theorem 1°) that if R^* is representable in $\mathcal{S}$ and $\mathcal{S}$ is consistent, then $\mathcal{S}$ is incomplete. We now show the following stronger theorem.

Theorem 1. *If some superset of R^* disjoint from P^* is representable in $\mathcal{S}$, then $\mathcal{S}$ is incomplete. More specifically, if $H(v_1)$ is a formula which represents in $\mathcal{S}$ some superset of R^* disjoint from P^*, then the sentence $H(\bar{h})$ is undecidable in $\mathcal{S}$, where h is the Gödel number of $H(v_1)$.*

Proof. Let A be the set represented by $H(v_1)$ and suppose

$$R^* \subseteq A$$

and that P^* is disjoint from A.

Since $H(v_1)$ represents A, then $H(\bar{h})$ is provable in $\mathcal{S}$ iff $h \in A$ (because for any number n, $H(\bar{n})$ is provable in $\mathcal{S}$ iff $n \in A$). Also $H(\bar{h})$ is provable in $\mathcal{S}$ iff $h \in P^*$ (by Lemma 1 of the last chapter). Hence $h \in P^*$ iff $h \in A$. But P^* is disjoint from A (by hypothesis) and, therefore, $h \notin P^*$ and $h \notin A$. Since $h \notin P^*$, then $H(\bar{h})$ is not provable in $\mathcal{S}$. Since $h \notin A$, then $h \notin R^*$ (because $R^* \subseteq A$) and so

$H(\bar{h})$ is not refutable in S. Therefore, $H(\bar{h})$ is undecidable in S.[1]

Remark. It was not necessary to add to the hypothesis of the above theorem that S is consistent because the hypothesis implies that P^* is disjoint from R^* and, hence, S must be consistent (because if S were inconsistent, then every formula would be provable, and h would be in both P^* and R^*).

The above theorem really is stronger than Theorem 1° of the last chapter because if S is consistent and if $H(v_1)$ represents R^* in S, then $H(v_1)$ certainly represents some superset of R^* disjoint from P^*—namely R^* itself.

Exercise 1. Show that if some superset of P^* disjoint from R^* is representable in S, then S is incomplete.

Separability in S. We shall say that a formula $F(v_1)$ *separates* a set A from a set B in S if for all $n \in A$, $F(\bar{n})$ is provable in S and for all $n \in B$, $F(\bar{n})$ is refutable in S.

Lemma 1. *If $F(v_1)$ separates A from B in S and S is consistent, then $F(v_1)$ represents some superset of A disjoint from B.*

Proof. Assume hypothesis. Let A' be the set represented by $F(v_1)$ in S. Since for all $n \in A$ the sentence $F(\bar{n})$ is provable, then $A \subseteq A'$. If some number n were in both A' and B, then $F(\bar{n})$ would be both provable and refutable in S (provable, since $F(v_1)$ represents A'). Hence if S is simply consistent, then A' is disjoint from B.

By the above lemma and Theorem 1 we at once have:

Theorem 2. *If $H(v_1)$ separates R^* from P^* in S and S is simply consistent, then $H(\bar{h})$ is undecidable in S, where h is the Gödel number of $H(v_1)$.*

Exercise 2. Show that if some formula separates P^* from R^* in S and S is consistent, then S is incomplete.

§2. A General Separation Principle.

We say that A is *separable* from B in S if there is a formula $F(v_1)$ which separates A from B in S. By virtue of Theorem 2, to show that P.A. is incomplete just on the basis of its simple consistency, it suffices to show that R^* is separable from P^* in P.A. This, in fact, is what Rosser

---

[1] This result is a special case of Exercise 3 of Chapter I.

did. As we pointed out in T.F.S., Rosser's method of separating R^* from P^* in P.A. easily generalizes to the separation of any two disjoint Σ_1-sets in P.A.—or indeed in any system S in which all true Σ_0-sentences are provable and in which all formulas of Ω_4 and Ω_5 are provable (in particular for S the system (Q) or even the system (R)). More generally, we say that a formula $F(v_1,\ldots,v_n)$ *separates* a relation $R_1(x_1,\ldots,x_n)$ from $R_2(x_1,\ldots,x_n)$ if for all numbers $k_1,\ldots,k_n$, if $R_1(k_1,\ldots,k_n)$ holds, then $F(\bar{k}_1,\ldots,\bar{k}_n)$ is provable in S, and if $R_2(k_1,\ldots,k_n)$ holds, then $F(\bar{k}_1,\ldots,\bar{k}_n)$ is refutable in S.

We call S a *Rosser system for sets* if for any Σ_1-sets A and B, the set $A - B$ is separable from $B - A$ in S. [This, of course, implies that for any *disjoint* Σ_1-sets A and B, the set A is separable from B in S]. And for each $n > 1$, we call S a *Rosser system for n-ary relations* if for any two Σ_1-relations

$$R_1(x_1,\ldots,x_n) \text{ and } R_2(x_1,\ldots,x_n),$$

the relation $R_1 - R_2$ (i.e., the relation $R_1(x_1,\ldots,x_n) \wedge \widetilde{R_2}(x_1,\ldots,x_n)$) is separable from $R_2 - R_1$ in S. We call S a *Rosser system* if S is a Rosser system for sets and for all n-ary relations. We will show that if all true Σ_0-sentences are provable in S and if all formulas of Ω_4 and Ω_5 are provable in S, then S is a Rosser system. More generally, we will show that if S is any system in which all formulas of Ω_4 and Ω_5 are provable, then for any two relations

$$R_1(x_1,\ldots,x_n) \text{ and } R_2(x_1,\ldots,x_n)$$

which are enumerable in S, their differences $R_1 - R_2$ and $R_2 - R_1$ are separable in S. We suggest the reader first try the following two exercises.

Exercise 3. Let $F_1(y)$ and $F_2(y)$ be formulas with y as the only free variable. Let n be any number.

1. Suppose the sentence $F_2(\bar{n})$ is true and for every number $m \leq n$, the sentence $F_1(\bar{m})$ is false. Is the sentence

$$\forall y(F_2(y) \supset (\exists z \leq y)F_1(z))$$

true or false?

2. Suppose the sentence $F_1(\bar{n})$ is true and for every number $m \leq n$, the sentence $F_2(\bar{m})$ is false. Is the sentence

$$\forall y(F_2(y) \supset (\exists z \leq y)F_1(z))$$

true or false?

A General Separation Principle

Exercise 4. Suppose all formulas of Ω_4 and Ω_5 are provable in S. Again let $F_1(y)$ and $F_2(y)$ be formulas with just y as a free variable, and let n be any number.

1. Suppose the sentence $F_1(\bar{n})$ is provable in S and that for every $m \leq n$, the sentence $F_2(\bar{m})$ is refutable in S. Show that the sentence
$$\forall y(F_2(y) \supset (\exists z \leq y)F_1(z))$$
is provable in S.

2. Suppose the sentence $F_2(\bar{n})$ is provable in S, and the sentence $F_1(\bar{m})$ is refutable in S, for every $m \leq n$. Show that the sentence
$$\forall y(F_2(y) \supset (\exists z \leq y)F_1(z))$$
is refutable in S.

The following lemma plays much the same role in Rosser's incompleteness proof that the ω-consistency lemma plays in Gödel's proof. It will have other applications as well.

Lemma S—Separation Lemma. If all formulas of Ω_4 and Ω_5 are provable in S, then for any two relations
$$R_1(x_1, \ldots x_n) \text{ and } R_2(x_1, \ldots, x_n)$$
enumerable in S, their differences $R_1 - R_2$ and $R_2 - R_1$ are separable in S.

Proof. Suppose all formulas of Ω_4 and Ω_5 are provable in S. We will show that for any two sets A and B enumerable in S, the set $B - A$ is separable from $A - B$ in S. (The proof for n-ary relations for $n > 1$ is an obvious modification which we leave to the reader.)

Let $A(x, y)$ and $B(x, y)$ be formulas that respectively enumerate A and B in S. We show that the formula
$$\forall y(A(x, y) \supset (\exists z \leq y)B(x, z))$$
separates $B - A$ from $A - B$ in S.

1. Suppose $n \in B - A$. Then $n \in B$, and for some k, the sentence $B(\bar{n}, \bar{k})$ is provable in S. Also, $n \notin A$, and so for every $m \leq k$ (in fact for every number m), the sentence $A(\bar{n}, \bar{m})$ is refutable, and by Ω_4, the sentence $(\forall y \leq \bar{k}) \sim A(\bar{n}, y)$ is provable. Hence the open formula $y \leq \bar{k} \supset \, \sim A(\bar{n}, y)$ is provable and, therefore, $A(\bar{n}, y) \supset \, \sim (y \leq \bar{k})$ is provable. By using Ω_5, it follows that $A(\bar{n}, y) \supset \bar{k} \leq y$ is provable, and since $B(\bar{n}, \bar{k})$ is provable, it

follows that $A(\bar{n},y) \supset (\bar{k} \leq y \wedge B(\bar{n},\bar{k}))$ is provable. Then (by first-order logic) the formula

$$A(\bar{n},y) \supset (\exists z \leq y)B(\bar{n},z)$$

is provable, hence so is the sentence

$$\forall y(A(\bar{n},y) \supset (\exists z \leq y)B(\bar{n},z)).$$

2. Suppose $n \in A - B$. Then for some k, $A(\bar{n},\bar{k})$ is provable and for all $m \leq k$ (in fact for all m), $B(\bar{n},\bar{m})$ is refutable and, therefore, (by Ω_4) the sentence $(\forall z \leq k) \sim B(\bar{n},z)$ is provable. It then follows that the sentence

$$A(\bar{n},\bar{k}) \wedge (\forall z \leq \bar{k}) \sim B(\bar{n},z)$$

is provable, hence so is the sentence

$$\sim (A(\bar{n},\bar{k}) \supset \sim (\forall z \leq \bar{k}) \sim B(\bar{n},z)),$$

and this is the sentence $\sim (A(\bar{n},\bar{k}) \supset (\exists z \leq \bar{k})B(\bar{n},z))$. Since the sentence $A(\bar{n},\bar{k}) \supset (\exists z \leq \bar{k})B(\bar{n},z)$ is refutable, so is the sentence $\forall y(A(\bar{n},y) \supset (\exists z \leq y)B(\bar{n},z))$.

Exercise 5. Derive Lemma S as a consequence of Exercise 4.

Exercise 6. Suppose all formulas of Ω_4 and Ω_5 are provable in $\mathcal{S}$, and suppose $A(x,y)$ and $B(x,y)$ respectively enumerate sets A and B in $\mathcal{S}$. Does the formula

$$\exists y(A(x,y) \wedge (\forall z \leq y)B(x,y))$$

separate $B - A$ from $A - B$ in $\mathcal{S}$? Does it separate $A - B$ from $B - A$ in $\mathcal{S}$?

It, of course, follows from the separation lemma that if all formulas of Ω_4 and Ω_5 are provable in $\mathcal{S}$, then for any *disjoint* sets A and B enumerable in $\mathcal{S}$ and B, it is separable from A in $\mathcal{S}$ (since then $B - A = B$ and $A - B = A$).

We showed in the last chapter that if all true Σ_0-sentences are provable in $\mathcal{S}$, then all Σ_1-relations are enumerable in $\mathcal{S}$. This with Lemma S gives:

Theorem 3. *Any extension of Ω_4 and Ω_5 in which all true Σ_0-sentences are provable is a Rosser system.*

Corollary. *The systems $(R), (Q)$ and P.A. are Rosser systems.*

§3. Rosser's Undecidable Sentence.

From Lemma S and Theorem 2 we get:

Theorem 4. *Let S be any simply consistent system in which the sets P^* and R^* are both enumerable and in which all formulas of Ω_4 and Ω_5 are provable. Then S is incomplete.*

Proof. Assume hypothesis. Then P^* and R^* are enumerable in S. By the assumption of simple consistency, the sets P^* and R^* must be disjoint. Then by Lemma S, R^* is separable from P^* in S. Hence S is incomplete by Theorem 2.

More specifically, suppose $A(x,y)$ is a formula that enumerates P^* in S, and $B(x,y)$ is a formula that enumerates R^* in S. As seen from the proof of Lemma S, the formula

$$\forall y(A(x,y) \supset (\exists z \leq y)B(x,z))$$

separates R^* from P^* in S. Then by Theorem 2, if h is the Gödel number of this formula, the sentence

$$\forall y(A(\bar{h},y) \supset (\exists z \leq y)B(\bar{h},z))$$

is undecidable in S (assuming S is simply consistent).

We can now easily prove Theorem R. Suppose S obeys the hypothesis of Theorem R. Since S is axiomatizable (by hypothesis), the sets P^* and R^* are both Σ_1. Then (by hypothesis), the sets P^* and R^* are both enumerable in S. Also (by hypothesis), S is an extension of Ω_4 and Ω_5. The conclusion then follows by Theorem 4.

More specifically, suppose S is an extension of Ω_4 and Ω_5, $A(x,y)$ is a formula that enumerates P^* in S, and $B(x,y)$ is a formula that enumerates R^* in S. Then by Lemma S, the formula

$$\forall y(A(x,y) \supset (\exists z \leq y)B(x,z))$$

separates $R^* - P^*$ from $P^* - R^*$. Assuming that S is consistent, the sets R^* and P^* are disjoint, and so the above formula separates R^* from P^*. We let h be the Gödel number of this formula. Then by Theorem 2, the sentence

$$\forall y(A(\bar{h},y) \supset (\exists z \leq y)B(\bar{h},z))$$

is undecidable in S (assuming that S is simply consistent).

The above sentence is Rosser's famous undecidable sentence. We now see how Peano Arithmetic can be shown incomplete just on the basis of its simple consistency (Gödel's stronger assumption of ω-consistency can be avoided).

§4. The Gödel and Rosser Sentences Compared.

We now consider the system P.A. Since P.A. is axiomatizable, the set P^* is Σ_1 and so there is a Σ_0-formula $A(x, y)$ that expresses a Σ_0-relation whose domain is P^*. Therefore, for any number n, $n \in P^*$ iff there is some number m such that $A(\bar{n}, \bar{m})$ is a *true* sentence. Also, $n \in P^*$ iff $E_n(\bar{n})$ is provable (in P.A.). Let us say that m is a *witness* that $E_n(\bar{n})$ is provable iff the sentence $A(\bar{n}, \bar{m})$ is true. Then $E_n(\bar{n})$ is provable iff there is a witness m that $E_n(\bar{n})$ is provable. Similarly, R^* is Σ_1 and, therefore, there is a Σ_0-formula $B(x, y)$ such that for any n, $E_n(\bar{n})$ is refutable iff there is some m such that $B(\bar{n}, \bar{m})$ is true; any such m we will call a *witness* that $E_n(\bar{n})$ is refutable.

Now, Gödel's sentence $\forall y \sim A(\bar{a}, y)$ (where a is the Gödel number of $\forall y \sim A(x, y)$) expresses the proposition that for all y, y is not a witness that $E_a(\bar{a})$ is provable, but $E_a(\bar{a})$ is the very sentence $\forall y \sim A(\bar{a}, y)$. And so the sentence can be read "For all y, y is not a witness that I am provable," or "there is no witness that I am provable," or more briefly: "I am not provable." Assuming ω-consistency, this sentence is undecidable in P.A.

The dual form of Gödel's sentence is $\exists y B(\bar{b}, y)$, where b is the Gödel number of the formula $\exists y B(x, y)$. This sentence can be thought of as saying: "There is a witness that I am refutable," or more briefly: "I am refutable." Again, under the assumption of ω-consistency, this sentence is undecidable in P.A.

Now let us consider Rosser's sentence

$$\forall y(A(\bar{h}, y) \supset (\exists z \leq y)B(\bar{h}, z)),$$

where h is the Gödel number of the formula

$$\forall y(A(x, y) \supset (\exists z \leq y)B(x, z)).$$

This sentence can be thought of as saying: "Given any witness y that I am provable, there is a number z less than or equal to y that is a witness that I am refutable."

Incidentally, the formula

$$\exists y(B(z, y) \wedge (\forall z \leq y) \sim A(x, z))$$

also separates R^* from P^* in P.A. (why?). If k is the Gödel number of this formula, then the sentence

$$\exists y(B(\bar{k}, y) \wedge (\forall z \leq y) \sim A(\bar{k}, z))$$

is also undecidable in P.A. (assuming P.A. is simply consistent). This sentence can be thought of as saying: "There is a witness that I am refutable and no number less than or equal to it is a witness that I am provable."

Exercise 7. The results of this exercise will be needed in the next chapter.

Suppose S is an extension of Ω_4 and Ω_5. Suppose also A is a set enumerable in S and $R(x,y)$ is a relation enumerable in S. Prove that the relation $x \in A \wedge \sim R(x,y)$ is separable in S from the relation $R(x,y) \wedge x \notin A$.

Exercise 8. Again let S be any system in which all formulas of Ω_4 and Ω_5 are provable, and let A and B be sets each of which is enumerable in S.

Prove that there is a formula $\psi(x,y)$ such that for any numbers n and m, the following two conditions hold.

1. If $n \in A$ and $m \notin B$, then $\psi(\bar{n}, \bar{m})$ is provable.
2. If $m \in B$ and $n \notin A$, then $\psi(\bar{n}, \bar{m})$ is refutable.

Exercise 9. The following proposition generalizes Lemma S and also directly provides solutions to the last two exercises.

Proposition S'. Let S be any system in which all formulas of Ω_4 and Ω_5 are provable. Let $R_1(x_1, \ldots, x_k)$ and $R_2(x_1, \ldots, x_n)$ be relations enumerable in S. Then there is a formula

$$\psi(x_1, \ldots, x_k, y_1, \ldots y_n)$$

with $x_1, \ldots x_k$ and $y_1, \ldots, y_n$ as the only free variables such that for all numbers $a_1, \ldots, a_k$ and $b_1, \ldots, b_n$, the following two conditions hold:

1. If $R_1(a_1, \ldots, a_k) \wedge \sim R_2(b_1, \ldots, b_n)$, then $\psi(\bar{a}_1, \ldots, \bar{a}_k, \bar{b}_1, \ldots, \bar{b}_n)$ is provable in S.
2. If $R_2(b_1, \ldots, b_n) \wedge \sim R_1(a_1, \ldots, a_k)$, then $\psi(\bar{a}_1, \ldots, \bar{a}_k, \bar{b}_1, \ldots, \bar{b}_n)$ is refutable in S.

(a) Prove Proposition S'.
(b) Show how Lemma S and the solutions of the last two exercises can be derived from Proposition S'.
(c) Show how Proposition S' can be derived from Exercise 4.

Exercise 10. Suppose S is a system (not necessarily one in which all formulas of Ω_4 and Ω_5 are provable) such that for any *disjoint* Σ_1-sets A and B, the set A is separable from B in S. Now suppose A and B are Σ_1-sets which are not disjoint. Is $A - B$ necessarily separable from $B - A$ in S?

§5. **More on Separation.** The solution of Exercise 10 is afforded by the following purely set-theoretic principle.

Consider two Σ_1-sets A and B. Then there are Σ_0-relations $R_1(x,y)$ and $R_2(x,y)$ with respective domains A and B. Let A' be the set of all n such that

$$\exists y (R_1(n,y) \wedge (\forall z \leq y) \sim R_2(n,z)),$$

and let B' be the set of all n such that

$$\exists y (R_2(n,y) \wedge (\forall z \leq y) \sim R_1(n,z)).$$

Let us say that y *puts* n in A iff $R_1(n,y)$ holds and that y puts n in B iff $R_2(n,y)$ holds. Then $n \in A$ iff some y puts n in A and $n \in B$ iff some y puts n in B. Let us say that n is put in A *before* n is put in B if there is some y that puts n in A and no $z \leq y$ puts n in B. The set A' is then the set of all n such that n is put in A before it is put in B and B' is the set of all n which is put in B before it is put in A. The sets A' and B' are clearly disjoint (since no number can be put in A before it is put in B and also be put in B before it is put in A). Also $A - B \subseteq A'$ because if $n \in A$ and $n \notin B$, then n is put in A before it is put in B, since n is not put in B at all. Finally, the sets A' and B' are both Σ_1 (since the relations

$$R_1(x,y) \wedge \forall z \leq y \sim R_2(x,z)$$

and

$$R_2(x,y) \wedge (\forall z \leq y) \sim R_1(x,z)$$

are Σ_0). This proves:

Theorem 5. *For any two Σ_1-sets A and B, there are disjoint Σ_1-sets A' and B' such that $(A - B) \subseteq A'$ and $(B - A) \subseteq B'$.*

Suppose now $\mathcal{S}$ is a system such that for any two *disjoint* Σ_1-sets A and B, the set A is separable from B in $\mathcal{S}$. Now, suppose A and B are Σ_1-sets not necessarily disjoint. Then by the above theorem, there are disjoint Σ_1-sets A' and B' such that $A - B \subseteq A'$ and $B - A \subseteq B'$. By hypothesis there is a formula $F(v_1)$ separating A' from B' in $\mathcal{S}$. Then $F(v_1)$ obviously separates $A - B$ from $B - A$ in $\mathcal{S}$ (because $n \in A - B \Rightarrow n \in A' \Rightarrow F(\bar{n})$ provable and $n \in B - A \Rightarrow n \in B' \Rightarrow F(\bar{n})$ refutable). And so we have:

Corollary. *If every disjoint pair of Σ_1-sets is separable in $\mathcal{S}$, then $\mathcal{S}$ is a Rosser system for sets.*

Of course, the above theorem and its corollary also hold for n-ary relations where $n > 1$. The reader can easily verify this.

Chapter VII

Shepherdson's Representation Theorems

We have already remarked that at the time of Gödel's proof, the only known way of showing the set P^* of Peano Arithmetic to be representable in P.A. involved the assumption of ω-consistency. Well, in 1960, A. Ehrenfeucht and S. Feferman showed that all Σ_1-sets can be represented in all *simply* consistent *axiomatizable* extensions of the system (R). Hence, all Σ_1-sets can be shown to be representable in P.A. under the weaker assumption that P.A. is simply consistent. Their proof combined a Rosser-type argument with a celebrated result in recursive function theory due to John Myhill which goes beyond the scope of this volume. Very shortly after, however, John Shepherdson [1961] found an extremely ingenious alternative proof that is more direct and which we study in this chapter. [In our sequel to this volume, we compare Shepherdson's proof with the original one. The comparison is of interest in that the two methods are very different and the proofs generalize in different directions which are apparently incomparable in strength.]

§1. **Shepherdson's Representation Theorem.** We recall that for each $n > 1$, a system S is called a Rosser system for n-ary relations if for any Σ_1-relations $R_1(x_1,\ldots,x_n)$ and $R_2(x_1,\ldots,x_n)$, the relation $R_1 - R_2$ is separable from $R_2 - R_1$ in S. We wish to prove the following theorem and its corollary (Th. 1 below).

Theorem S_1—Shepherdson's Representation Theorem. *If S is a simply consistent axiomatizable Rosser system for binary relations (n-ary relations for $n = 2$), then all Σ_1-sets are representable in S.*

Theorem 1—Ehrenfeucht, Feferman. *All Σ_1-sets are representable in every consistent axiomatizable extension of the system (R).*

Shepherdson's Lemma and Weak Separation. For emphasis, we will now sometimes write "strongly separates" for "separates". We will say that a formula $F(v_1)$ *weakly* separates A from B in S if $F(v_1)$ represents some superset of A disjoint from B. We showed in the last chapter (Lemma 1) that strong separation implies weak separation provided that the system S is consistent. We also say that a formula $F(v_1,\ldots,v_n)$ weakly separates a relation $R_1(x_1,\ldots,x_n)$ from $R_2(x_1,\ldots,x_n)$ if $F(v_1,\ldots,v_n)$ represents some relation $R'(x_1,\ldots,x_n)$ such that $R_1 \subseteq R_1'$ and R_1' is disjoint from R_2. With relations as with sets, strong separation implies weak separation if the system S is consistent (as the reader can easily verify). Let us note that to say that $F(v_1,\ldots,v_n)$ weakly separates $R_1(x_1,\ldots,x_n)$ from $R_2(x_1,\ldots,x_n)$ is equivalent to saying that for every n-tuple $(k_1,\ldots,k_n) \in R_1$, the sentence $F(\overline{k}_1,\ldots,\overline{k}_n)$ is provable, and for every n-tuple $(k_1,\ldots,k_n) \in R_2$, the sentence $F(\overline{k}_1,\ldots,\overline{k}_n)$ is not provable.

The Function $\Pi(x,y,z)$. For any expression E, we let $E[\overline{m},\overline{n}]$ be the expression

$$\forall v_2(v_2 = \overline{n} \supset \forall v_1(v_1 = \overline{m} \supset E)).$$

If E is a formula in which v_1 and v_2 are the only free variables, then $E[\overline{m},\overline{n}]$ is a sentence equivalent to the sentence $E(\overline{m},\overline{n})$ (by which we mean the result of substituting $\overline{m}$ and $\overline{n}$ for all free occurrences of v_1 and v_2 respectively in E). Also, if E is a formula, then the formula

$$E[\overline{m},\overline{n}] \equiv E(\overline{m},\overline{n})$$

is logically valid. Hence it is provable in S and, therefore, $E[\overline{m},\overline{n}]$ is provable in S iff $E(\overline{m},\overline{n})$ is provable in S.

We now define $\Pi(x,y,z)$ to be the Gödel number of $E_x[\overline{y},\overline{z}]$ (E_x is the expression whose Gödel number is x). We note that if

$$\Pi(x,y,z) \in P,$$

then E_x is automatically a formula. Hence $E_x(\overline{y},\overline{z})$ is provable (since $E_x[\overline{y},\overline{z}]$ is provable). Conversely, if $E_x(\overline{y},\overline{z})$ is provable, then so is $E_x[\overline{y},\overline{z}]$, and hence $\Pi(x,y,z) \in P$. So for any numbers x, y and z, $E_x[\overline{y},\overline{z}]$ is provable $\leftrightarrow$ $E_x(\overline{y},\overline{z})$ is provable $\leftrightarrow$ $\Pi(x,y,z) \in P$.

Lemma 1—Shepherdson's Representation Lemma. *For any set A, if the relation $x \in A \wedge \Pi(y,x,y) \notin P$ is weakly separable in S from the relation $\Pi(y,x,y) \in P \wedge x \notin A$, then A is representable in S. More specifically, if $E_h(v_1,v_2)$ is a formula that effects the separation, then $E_h(v_1,\overline{h})$ represents A in S.*

The above lemma easily follows from the following lemma, which will have other applications later on.

Lemma 1*. *For any relation $R(x,y)$, if $E_h(v_1,v_2)$ is a formula that weakly separates the relation*

$$R(x,y) \wedge \Pi(y,x,y) \notin P$$

from

$$\Pi(y,x,y) \in P \wedge {\sim} R(x,y),$$

then for every number n, the sentence $E_h(\overline{n},\overline{h})$ is provable in S iff $R(n,h)$ holds.

Proof. Assume hypothesis. Then for any numbers n and m:

(1) $R(n,m) \wedge \Pi(m,n,m) \notin P \Rightarrow E_h(\overline{n},\overline{m})$ is provable.
(2) $\Pi(m,n,m) \in P \wedge {\sim} R(n,m) \Rightarrow E_h(\overline{n},\overline{m})$ is not provable.

By (1), taking h for m, if $R(n,h)$ and $\Pi(h,n,h) \notin P$, then $E_h(\overline{n},\overline{h})$ is provable. This means that if $R(n,h)$ and $E_h(\overline{n},\overline{h})$ is not provable, then $E_h(\overline{n},\overline{h})$ is provable, from which it follows that if $R(n,h)$, then $E_h(\overline{n},\overline{h})$ is provable.

By (2), taking h for m, if $\Pi(h,n,h) \in P$ and ${\sim} R(n,h)$, then $E_h(\overline{n},\overline{h})$ is not provable, which means that if $E_h(\overline{n},\overline{h})$ is provable and ${\sim} R(n,h)$, then $E_h(\overline{n},\overline{h})$ is not provable, from which follows that if $E_h(\overline{n},\overline{h})$ is provable, then $R(n,h)$ must hold. Therefore, (by (1)) $E_h(\overline{n},\overline{h})$ is provable iff $R(n,h)$.

Lemma 1 results from Lemma 1* by taking for R the set of all ordered pairs $\langle x,y\rangle$ such that $x \in A$. If the hypothesis of Lemma 1 holds, then by Lemma 1*, for all n, $E_h(\overline{n},\overline{h})$ is provable $\leftrightarrow R(n,h) \leftrightarrow n \in A$. Therefore, the formula $E_h(v_1,\overline{h})$ represents A in S.

Now we can easily prove Theorem S_1. Using the fact that the relation $x *_{13} y = z$ and the relation $13^x = y$ are both Σ_1, the relation $\Pi(x,y,z) = w$ is easily seen to be Σ_1. Therefore, the relation $\Pi(y,x,y) = z$ (as a relation among x,y and z) is Σ_1, and so if S is axiomatizable, then the set P is Σ_1. Hence the relation

$$\Pi(y,x,y) \in P$$

is Σ_1 (it can be written as $\exists z(\Pi(y,x,y) = z \wedge z \in P)$). For any Σ_1-set A, let $R_A(x,y)$ be the relation $x \in A$. The relation $R_A(x,y)$ is also Σ_1 (it can be written as $x \in A \wedge y = y$). Then by the hypothesis of Theorem S_1, the relation

$$R_A(x,y) \wedge \sim \Pi(y,x,y) \in P$$

is strongly separable in $\mathcal{S}$ from the relation

$$\Pi(y,x,y) \in P \wedge \sim R_A(x,y).$$

Thus, the relation

$$x \in A \wedge \sim \Pi(y,x,y) \in P$$

is strongly separable from the relation

$$\Pi(y,x,y) \in P \wedge \sim x \in A.$$

By the assumption of simple consistency, the first of the above relations is weakly separable from the second. Then by Lemma 1, A is representable in $\mathcal{S}$. This proves Theorem S_1.

Suppose now $\mathcal{S}$ is an axiomatizable consistent extension of (R) (or for that matter, any consistent axiomatizable extension of Ω_4 and Ω_5 in which all true Σ_0-sentences are provable). If A is a Σ_1-set, then there is a Σ_0-formula $A(x,y)$ that enumerates the set A in $\mathcal{S}$. The relation $\Pi(y,x,y) \in P$ is also Σ_1, so there is a Σ_0-formula $B(x,y,z)$ that enumerates this relation in $\mathcal{S}$. Then the formula

$$\forall z(B(x,y,z) \supset (\exists w \leq z)A(x,w))$$

strongly separates (hence also weakly separates, since $\mathcal{S}$ is consistent) the relation

$$x \in A \wedge \Pi(y,x,y) \notin P$$

from

$$\Pi(y,x,y) \in P \wedge x \notin A$$

in $\mathcal{S}$. Letting h be the Gödel number of this formula, the formula

$$\forall z(B(x,\overline{h},z) \supset (\exists w \leq z)A(x,w))$$

represents A in $\mathcal{S}$, by Lemma 1. We, thus, have a concrete idea of what the representing formula for A looks like.

Discussion. Let us now take a closer look into the significance of the above formula. Let us say that z is a *witness* that x is in A if the Σ_0-sentence $A(\overline{x}, \overline{z})$ is true. Let us say that z is a *witness* that $E_y(\overline{x}, \overline{y})$ is provable if the Σ_0-sentence $B(\overline{x}, \overline{y}, \overline{z})$ is true. Then for any number n, the sentence

$$\forall z(B(\overline{n}, \overline{h}, z) \supset (\exists w \leq z)A(\overline{n}, w))$$

can be read, "For any witness z that $E_h(\overline{n}, \overline{h})$ is provable, there is a witness $w \leq z$ that n is in A." However, $E_h(\overline{n}, \overline{h})$ is this very sentence! And so the sentence can be paraphrased, "Given any witness that I am provable, some number less than or equal to it is a witness that $n \in A$". The sentence, therefore, is self-referential; it refers not only to n's membership in A, but also to its own provability in $\mathcal{S}$.

An alternative formula that works is the formula

$$\exists z(A(x, z) \wedge (\forall w \leq z) \sim B(x, \overline{k}, w)),$$

where k is the Gödel number of

$$\exists z(A(x, z) \wedge (\forall w \leq z) \sim B(x, y, w)).$$

For any n, the sentence

$$\exists z(A(\overline{n}, z) \wedge (\forall x \leq z) \sim B(\overline{n}, \overline{k}, w))$$

can be read: "There is a witness that n is in A, and no number less than or equal to it is a witness that I am provable."

Exercise 1. Show that for every positive n, if $\mathcal{S}$ is a consistent axiomatizable Rosser system for $(n + 1)$-ary relations, then every n-ary Σ_1-relation $R(x_1, \ldots, x_n)$ is representable in $\mathcal{S}$.

§2. Exact Rosser Systems.

For any disjoint pair (A, B) of sets, we say that a formula $F(v_1)$ *exactly* separates A from B in $\mathcal{S}$ (or that $F(v_1)$ exactly separates the ordered pair (A, B) in $\mathcal{S}$) if $F(v_1)$ represents A and its negation $\sim F(v_1)$ represents B in $\mathcal{S}$. This means that for all $n \in A$, $F(\overline{n})$ is provable; for all $n \in B$, $F(\overline{n})$ is refutable, and for all $n \notin (A \cup B)$, $F(\overline{n})$ is undecidable in $\mathcal{S}$. [We note that if A is exactly separable from B in $\mathcal{S}$, it is also strongly and weakly separable from B, and $\mathcal{S}$ must be simply consistent.]

We call $\mathcal{S}$ an *exact* Rosser system for sets if every disjoint pair (A, B) of Σ_1-sets is exactly separable in $\mathcal{S}$. In the Putnam-Smullyan paper [1960], we proved that every consistent axiomatizable extension of (R) is an exact Rosser system for sets (in fact for n-ary relations for every n). The argument used a result of Smullyan from recursion theory (a double analogue of Myhill's theorem) that we will give in our sequel. But Shepherdson gave an alternative proof that "doubles up" on his construction of the last section. Again, his method and the method of Putnam-Smullyan generalize in different directions which we will compare in our sequel.

Informally, what Shepherdson did is this. Given two formulas $A(x,y)$ and $B(x,y)$, enumerating two disjoint Σ_1-sets A and B in S where S is an axiomatizable extension of (R), Shepherdson constructed a formula $\phi(x)$ such that for any number n, the sentence $\phi(\overline{n})$ expresses the proposition, "For any z, if z is either a witness that I am provable or a witness that n is in B, then there exists some $w \leq z$ that is either a witness that I am refutable or a witness that n is in A." Now for the formal details. [We recall that R is the set of Gödel numbers of the refutable formulas.]

Lemma 2—Shepherdson's Separation Lemma. *For any disjoint sets A and B, if the relation*

$$(x \in A \vee \Pi(y,x,y) \in R) \wedge \sim (x \in B \vee \Pi(y,x,y) \in P)$$

is strongly separable from

$$(x \in B \vee \Pi(y,x,y) \in P) \wedge \sim (x \in A \vee \Pi(y,x,y) \in R)$$

and if S is simply consistent, then A is exactly separable from B in S.

We shall prove the following strengthening of Lemma 2 (which will have an application in our sequel).

Lemma 2*. *Suppose $R_1(x,y)$ and $R_2(x,y)$ are disjoint relations. Let $S_1(x,y)$ be the relation*

$$R_1(x,y) \vee \Pi(y,x,y) \in R,$$

and let $S_2(x,y)$ be the relation

$$R_2(x,y) \vee \Pi(y,x,y) \in P.$$

Then if $E_h(v_1,v_2)$ is a formula that strongly separates $S_1 - S_2$ from $S_2 - S_1$ in S and S is consistent, then for any number n:

1. *$R_1(n,h)$ iff $E_h(\overline{n},\overline{h})$ is provable in S.*
2. *$R_2(n,h)$ iff $E_h(\overline{n},\overline{h})$ is refutable in S.*

Proof. Assume hypothesis. Taking h for y, we have for all n:

(1) $[R_1(n,h) \lor \Pi(h,n,h) \in R] \land \sim [R_2(n,h) \lor \Pi(h,n,h) \in P] \Rightarrow E_h(\overline{n},\overline{h})$ is provable.
(2) $[R_2(n,h) \lor \Pi(h,n,h) \in P] \land \sim [R_1(n,h) \lor \Pi(h,n,h) \in R] \Rightarrow E_h(\overline{n},\overline{h})$ is refutable.

    (a) Now suppose $R_1(n,h)$. Then the left conjunctive clause of the antecedent of (1) is true and the right conjunctive clause reduces to $\Pi(h,n,h) \notin P$ (because R_1 is disjoint from R_2, so $R_2(n,h)$ is false). Therefore, (1) reduces to $\Pi(h,n,h) \notin P \Rightarrow E_h(\overline{n},\overline{h})$ is provable, which means that if $E_h(\overline{n},\overline{h})$ is not provable, then it is provable. Hence $E_h(\overline{n},\overline{h})$ must be provable (assuming $R_1(n,h)$). Conversely, suppose $E_h(\overline{n},\overline{h})$ is provable. Then

$$\Pi(h,n,h) \in P,$$

so the left conjunctive clause of the antecedent of (2) is true and the right conjunctive clause reduces to

$$\sim R_1(n,h)$$

(because $\Pi(h,n,h) \in R$ is false, since $\Pi(h,n,h) \in P$ and S is assumed consistent). Therefore, $\sim R_1(n,h) \Rightarrow E_h(\overline{n},\overline{h})$ is refutable. But $E_h(\overline{n},\overline{h})$ is not refutable (by consistency) and so $R_1(n,h)$ holds. This proves (1).
    (b) The proof of (2) is symmetric and is left to the reader.

Remark. The above proof embodies a cute principle of propositional logic. Suppose we have four propositions r_1, r_2, q_1 and q_2, and suppose the following four conditions are given:

(1) $[(r_1 \lor q_2) \land \sim (r_2 \lor q_1)] \Rightarrow q_1$
(2) $[(r_2 \lor q_1) \land \sim (r_1 \lor q_2)] \Rightarrow q_2$
(3) $\sim (r_1 \land r_2)$
(4) $\sim (q_1 \land q_2)$

It then follows that $r_1 \leftrightarrow q_1$ and $r_2 \leftrightarrow q_2$.

Coming back to our subject, Lemma 2 is, of course, a consequence of Lemma 2* (define $R_1(x,y)$ iff $x \in A$ and $R_2(x,y)$ iff $x \in B$), and from Lemma 2 we get:

Theorem S_2—Shepherdson's Separation Theorem. *Every consistent axiomatizable Rosser system for binary relations is an exact Rosser system for sets.*

Proof. If S is axiomatible, then the relations $\Pi(y,x,y) \in P$ and $\Pi(y,x,y) \in R$ are both Σ_1. Hence for any Σ_1-sets A and B, the relation

$$x \in A \vee \Pi(y,x,y) \in R$$

and the relation

$$x \in B \vee \Pi(y,x,y) \in P$$

are both Σ_1. If, furthermore, S is a Rosser system for binary relations, then the differences of the above Σ_1-relations are strongly separable in S. Hence the conclusion follows by the above lemma.

Theorem 2—Putnam-Smullyan. *Every simply consistent axiomatizable extension of (R) is an exact Rosser system for sets.*

Proof. By Theorem S_2 and the fact that every extension of (R) is a Rosser system.

Exercise 2. Suppose S is a consistent axiomatizable extension of (R). Suppose also A and B are disjoint Σ_1-sets, and $A(x,y)$ and $B(x,y)$ are Σ_0-formulas that respectively enumerate A and B in S. Suppose $C(x,y,z)$ and $D(x,y,z)$ are formulas that respectively enumerate the relations

$$\Pi(y,x,y) \in P$$

and

$$\Pi(y,x,y) \in R.$$

Construct a formula that exactly separates A from B in S. [This can be done in two ways—one involves a formula beginning with a universal quantifier and the other involves a formula beginning with an existential quantifier.]

§3. Some Variants of Rosser's Undecidable Sentence.

Shepherdson's methods have suggested to us some curious variants of Rosser's undecidable sentence.

Suppose S is a simply consistent system, not necessarily axiomatizable. We know that if either of the sets P^* and R^* is enumerable in S and S is ω-consistent, then S is incomplete. We also know that if *both* the sets P^* and R^* are enumerable in S and S is an extension of Ω_4 and Ω_5, then S is incomplete (by Rosser's argument). Our present point is that in place of the two sets P^* and R^*, we can get by with the single triadic relation $\Pi(x,y,z) \in P$—its enumerability in S is enough to ensure incompleteness (assuming S is a consistent extension of Ω_4 and Ω_5). In fact, we can even get by with the single binary relation $\Pi(y,x,y) \in P$. We will show these facts by using some variants of Shepherdson's arguments.

In that which follows, S is assumed simply consistent, but not necessarily axiomatizable.

Theorem 3. *If the relation*

$$\Pi(x,x,y) \in P \vee \Pi(y,x,y) \notin P$$

is weakly separable in S from the relation

$$\Pi(y,x,y) \in P \vee \Pi(x,x,y) \notin P,$$

then S is inconsistent or incomplete.

Proof. Let $E_h(v_1, v_2)$ be a formula which effects the separation. Then by Lemma 1*, for any n,

$$\Pi(n,n,h) \in P \leftrightarrow E_h(\overline{n}, \overline{h})$$

is provable. Hence $E_n(\overline{n}, \overline{h})$ is provable iff $E_h(\overline{n}, \overline{h})$ is provable. Now let $E_k(v_1, v_2)$ be the negation of $E_h(v_1, v_2)$. Then $E_k(\overline{k}, \overline{h})$ is provable iff $E_h(\overline{k}, \overline{h})$ is refutable. But by the preceding sentence, it is also true that $E_k(\overline{k}, \overline{h})$ is provable $\leftrightarrow E_h(\overline{k}, \overline{h})$ is provable. This means that $E_h(\overline{k}, \overline{h})$ is refutable $\leftrightarrow E_h(\overline{k}, \overline{h})$ is provable. Hence S is inconsistent or incomplete. Under the assumption that S is consistent, the sentence $E_h(\overline{k}, \overline{h})$ is undecidable in S.

Corollary. *If the relation $\Pi(x,y,z) \in P$ is enumerable in S and S is a consistent extension of Ω_4 and Ω_5, then S is incomplete.*

Proof. Suppose $F(v_1, v_2, v_3, v_4)$ is a formula that enumerates the relation $\Pi(x,y,z) \in P$. Then obviously $F(v_1, v_1, v_2, v_3)$ enumerates the relation $\Pi(x,x,y) \in P$ and $F(v_2, v_1, v_2, v_3)$ enumerates the relation $\Pi(y,x,y) \in P$. Therefore, the hypothesis implies that the relations $\Pi(x,x,y) \in P$ and $\Pi(y,x,y) \in P$ are both enumerable in S. Hence their differences (in either order) are strongly separable in S, and so the conclusion follows from Theorem 3.

More specifically, it can be easily seen that if h is the Gödel number of the formula

$$\forall v_3(F(v_2, v_1, v_2, v_3) \supset (\exists v_4 \leq v_3) F(v_1, v_1, v_2, v_4))$$

and k is the Gödel number of its negation, then the sentence
$$\forall v_3(F(\overline{h},\overline{k},\overline{h},v_3) \supset (\exists v_4 \le v_3)F(\overline{k},\overline{k},\overline{h},v_4))$$
is undecidable in S (if S is consistent).

We have remarked that we could use the relation $\Pi(y,x,y) \in P$ in place of the relation $\Pi(x,y,z) \in P$. This can be done in at least two different ways, indicated in the exercises that follow.

Exercise 4. Suppose there is a number h such that for all n, $E_h(\overline{n},\overline{h})$ is provable in S iff $E_n(\overline{h},\overline{n})$ is provable in S. Show that S, if consistent, is incomplete. [Hint: Consider the Gödel number k, not of the formula $\sim E_h(v_1,v_2)$, but of the formula $\sim E_h(v_2,v_1)$.]

Exercise 5. Using Exercise 4 and Lemma 1*, prove that if the relation
$$\Pi(x,y,x) \in P \wedge \Pi(y,x,y) \notin P$$
is weakly separable in S from the relation
$$\Pi(y,x,y) \in P \wedge \Pi(x,y,x) \notin P.$$
Then S, if consistent, is incomplete.

Exercise 6. Using Exercise 5, show that if the relation
$$\Pi(y,x,y) \in P$$
is enumerable in S and S is a consistent extension of Ω_4 and Ω_5, then S is incomplete.

Exercise 7. How does Exercise 6 yield another proof of Theorem 3?

Exercise 8. Show that if the set of x such that $\Pi(x,x,x) \in P$ is representable in S and S is consistent, then S is incomplete.

Exercise 9. Using Exercise 8, show that if the relation
$$\Pi(x,x,x) \in P \wedge \Pi(y,x,y) \notin P$$
is weakly separable in S from the relation
$$\Pi(y,x,y) \in P \wedge \Pi(x,x,x) \notin P,$$
then S, if consistent, is incomplete.

Exercise 10. Show that Exercise 9 provides another solution of Exercise 6.

§4. A Strengthening of Shepherdson's Theorems.

If we use Lemmata 1* and 2* in place of Lemmata 1 and 2 respectively, we get the following strengthening of Theorems S_1 and S_2:

Theorem S_1^*. *If S is a consistent axiomatizable Rosser system for binary relations, then for any Σ_1-relation $R(x, y)$, there is a number h such that $E_h(v_1, v_2)$ is a formula, and $E_h(v_1, \overline{h})$ represents the set of all n such that $R(n, h)$.*

Theorem S_2^*. *Under the same hypothesis, for any disjoint Σ_1-relations $R_1(x, y)$ and $R_2(x, y)$, there is a number h such that $E_h(v_1, v_2)$ is a formula, and $E_h(v_1, \overline{h})$ exactly separates the set of all n such that $R_1(n, h)$ from the set of all n such that $R_2(n, h)$.*

These results will have interesting applications in our sequel. Their proofs are obvious modifications of the proofs of Theorems S_1 and S_2.

Exercise 11. Prove Theorems S_1^* and S_2^*.

Chapter VIII

Definability and Diagonalization

In this chapter we establish some basic facts about Σ_1-relations and functions that will be needed for the rest of this study. We also introduce the notion of *fixed-points* of formulas and prove a fundamental fact about them which is crucial for Gödel's *second* incompleteness theorem and related results of the next chapter.

§1. Definability and Complete Representability.

A formula $F(v_1,\ldots,v_n)$ is said to *define* a relation $R(x_1,\ldots,x_n)$ in a system $\mathcal{S}$ if for all numbers $a_1,\ldots,a_n$, the two following conditions hold.

(1) $R(a_1,\ldots,a_n) \Rightarrow F(\bar{a}_1,\ldots,\bar{a}_n)$ is provable in $\mathcal{S}$.
(2) $\tilde{R}(a_1,\ldots,a_n) \Rightarrow F(\bar{a}_1,\ldots,\bar{a}_n)$ is refutable in $\mathcal{S}$.

We say that $F(v_1,\ldots,v_n)$ *completely represents* $R(x_1,\ldots,x_n)$ in $\mathcal{S}$ iff F represents R and $\sim F$ represents the complement $\tilde{R}$ of R in $\mathcal{S}$—in other words, if (1) and (2) above hold with "$\Rightarrow$" replaced by "$\leftrightarrow$".

Proposition 1. *If F defines R in $\mathcal{S}$ and $\mathcal{S}$ is consistent, then F completely represents R in $\mathcal{S}$.*

Proof. Assume hypothesis. We must show that the converses of (1) and (2) above must hold.

Suppose $F(\bar{a}_1,\ldots,\bar{a}_n)$ is provable in $\mathcal{S}$. Then $F(\bar{a}_1,\ldots,\bar{a}_n)$ is not refutable in $\mathcal{S}$ (by the assumption of consistency). Therefore by (2), $\tilde{R}(a_1,\ldots,a_n)$ cannot hold. Hence $R(a_1,\ldots,a_n)$ holds.

Similarly, if $F(\bar{a}_1,\ldots,\bar{a}_n)$ is refutable, then it is not provable. Hence by (1), $R(a_1,\ldots,a_n)$ cannot hold and hence $\tilde{R}(a_1,\ldots,a_n)$.

Recursive Relations. By a *recursive* set or relation, we mean one such that it and its complement are both Σ_1. [There are many different, but equivalent, definitions in the literature of *recursive* relations. We will consider some others in the sequel to this volume.]

It is obvious that a formula F defines a relation R in S iff F separates R from $\tilde{R}$ in S. Suppose now S is a Rosser system and that R is a recursive relation. Then R and $\tilde{R}$ are both Σ_1. Hence R is separable from $\tilde{R}$ in S, which means that R is definable in S. And so we have:

Proposition 2.

1. *If S is a Rosser system, then all recursive relations are definable in S.*
2. *If S is a consistent Rosser system, then all recursive relations are completely representable in S.*

Statement (2) follows from (1) by Proposition 1. In the last chapter we proved that the system (R) is a Rosser system and so we have:

Theorem 1. *All recursive relations are definable in (R).*

Note. If a set or relation is definable in a system S, it is obviously definable in every extension of S. Therefore, all recursive relations are definable in all consistent extensions of (R)—in particular, in the systems (Q) and P.A.

Exercise 1. Show that if $F(v_1, v_2)$ defines $R(x_1, x_2)$ in S and A is the domain of R, then $F(v_1, v_2)$ enumerates A in S.

Exercise 2. State whether the following is true or false: If all true Σ_0-sentences are provable in S, then all Σ_0-relations are definable in S.

Exercise 3. Show that for the complete theory $\mathcal{N}$, representability, definability and complete representability all coincide. Is this true for P.A. rather than $\mathcal{N}$? [Hint: Is the set P^* completely representable in P.A.?]

§2. Strong Definability of Functions in S. A formula

$$F(v_1, \ldots, v_n, v_{n+1})$$

will be said to *weakly* define a function $f(x_1, \ldots, x_n)$ in S if it defines the relation

$$f(x_1, \ldots, x_n) = x_{n+1}$$

in S. We shall say that the formula *strongly* defines, or more briefly, *defines* the function $f(x_1, \ldots, x_n)$ in S iff for all numbers $a_1, \ldots, a_n$ and b, the following three conditions hold:

(1) If $f(a_1, \ldots, a_n) = b$, then $F(\bar{a}_1, \ldots, \bar{a}_n, \bar{b})$ is provable in S.
(2) If $f(a_1, \ldots, a_n) \neq b$, then $F(\bar{a}_1, \ldots, \bar{a}_n, \bar{b})$ is refutable in S.
(3) If $f(a_1, \ldots, a_n) = b$, then the sentence

$$\forall v_{n+1}(F(\bar{a}_1, \ldots, \bar{a}_n, v_{n+1}) \supset v_{n+1} = \bar{b})$$

is provable in S.

Conditions (1) and (2) jointly say that F weakly defines f in S. Thus, F strongly defines f in S iff F weakly defines f in S and condition (3) holds.

We will be concerned mainly with functions of one argument. The next theorem and its corollaries reveal the significance of strong definability.

Theorem 2. *If $f(x)$ is strongly definable in S, then for any formula $G(v_1)$, there is a formula $H(v_1)$ such that for any number n, the sentence $H(\bar{n}) \equiv G(\overline{f(n)})$ is provable in S.*

Proof. Suppose $F(v_1, v_2)$ strongly defines $f(x)$ in S. Given a formula $G(v_1)$, we let $H(v_1)$ be the formula

$$\exists v_2(F(v_1, v_2) \wedge G(v_2)).$$

We show that the formula $H(v_1)$ works.

Take any n and m such that $f(n) = m$. We are to show that the sentence $H(\bar{n}) \equiv G(\bar{m})$ is provable in S.

1. Since $F(v_1, v_2)$ defines $f(x)$, then $F(\bar{n}, \bar{m})$ is provable in S. Hence $G(\bar{m}) \supset (F(\bar{n}, \bar{m}) \wedge G(\bar{m}))$ is provable and, therefore, $G(\bar{m}) \supset \exists v_2(F(\bar{n}, v_2) \wedge G(v_2))$ and $G(\bar{m}) \supset H(\bar{n})$ are provable in S.

2. It follows from condition (3) of strong definability that the open formula $F(\bar{n}, v_2) \supset v_2 = \bar{m}$ is provable. Therefore,

$$(F(\bar{n}, v_2) \wedge G(v_2)) \supset (v_2 = \bar{m} \wedge G(v_2))$$

is provable, and $(v_2 = \bar{m} \wedge G(v_2)) \supset G(\bar{m})$ is logically valid. Hence provable in S. Therefore, $(F(\bar{n}, v_2) \wedge G(v_2)) \supset G(\bar{m})$ is

provable. Hence by first-order logic,
$$\exists v_2(F(\overline{n}, v_2) \wedge G(v_2)) \supset G(\overline{m})$$
is provable—i.e. $H(\overline{n}) \supset G(\overline{m})$ is provable.

By (1) and (2), the sentence $H(\overline{n}) \equiv G(\overline{m})$ is provable in $\mathcal{S}$.

Corollary. *Suppose $f(x)$ is strongly definable in $\mathcal{S}$. Then*

1. *For any set A representable in $\mathcal{S}$, the set $f^{-1}(A)$ is representable in $\mathcal{S}$.*
2. *For any pair (A, B) that is exactly separable in $\mathcal{S}$, the pair $(f^{-1}(A), f^{-1}(B))$ is exactly separable in $\mathcal{S}$.*
3. *For any set A definable in $\mathcal{S}$, the set $f^{-1}(A)$ is definable in $\mathcal{S}$.*

Proof. Suppose $f(x)$ is strongly definable in $\mathcal{S}$. Then by Theorem 2, for any formula $G(v_1)$, there is a formula $H(v_1)$ such that for any n, the sentence $H(\overline{n}) \equiv G(\overline{f(n)})$ is provable in $\mathcal{S}$. This, of course, implies that for any n, $H(\overline{n})$ is provable in $\mathcal{S}$ iff $G(\overline{f(n)})$ is provable in $\mathcal{S}$ and $H(\overline{n})$ is refutable in $\mathcal{S}$ iff $G(\overline{f(n)})$ is refutable in $\mathcal{S}$.

1. Suppose $G(v_1)$ represents A in $\mathcal{S}$. Then for any n, $n \in f^{-1}(A) \leftrightarrow f(n) \in A \leftrightarrow G(\overline{f(n)})$ is provable in $\mathcal{S} \leftrightarrow H(\overline{n})$ is provable in $\mathcal{S}$. Hence $H(v_1)$ represents $f^{-1}(A)$ in $\mathcal{S}$.
2. Suppose also the negation of $G(v_1)$ represents B. Then for any n, $n \in f^{-1}(B) \leftrightarrow f(n) \in B \leftrightarrow G(\overline{f(n)})$ is refutable in $\mathcal{S} \leftrightarrow H(\overline{n})$ is refutable in $\mathcal{S}$. Therefore, $\sim H(v_1)$ represents $f^{-1}(B)$. So $H(v_1)$ exactly separates $(f^{-1}(A), f^{-1}(B))$ in $\mathcal{S}$.
3. This follows from 2, taking A for B.

Exercise 4. Suppose $f(x)$ is weakly definable in $\mathcal{S}$ without necessarily being strongly definable in $\mathcal{S}$. Show that if $\mathcal{S}$ is ω-consistent, then for any set A definable in $\mathcal{S}$, the set $f^{-1}(A)$ is representable in $\mathcal{S}$.

§3. Strong Definability of Recursive Functions in (R).

A function $f(x_1, \ldots, x_n)$ is called recursive iff the relation
$$f(x_1, \ldots, x_n) = x_{n+1}$$
is recursive. By Theorem 1, all recursive functions are weakly definable in (R). We now wish to prove

Theorem 3. *All recursive functions are strongly definable in (R).*

This will follow from Theorem 1 once we have proved:

Lemma. *If all formulas of Ω_4 and Ω_5 are provable in S, then any function weakly definable in S is strongly definable in S.*

Proof. We illustrate the proof for functions of one argument.

Suppose all formulas of Ω_4 and Ω_5 are provable in S and $F(x,y)$ is a formula that weakly defines $f(x)$ in S. Let $G(x,y)$ be the formula

$$F(x,y) \wedge \forall z(F(x,z) \supset y \leq z).$$

We show that $G(x,y)$ strongly defines $f(x)$ in S.

Suppose $f(n) = m$. We are to show three things:

(1) $G(\bar{n}, \bar{m})$ is provable in S.
(2) For every $k \neq m$, $G(\bar{n}, \bar{k})$ is refutable in S.
(3) $\forall y(G(\bar{n}, y) \supset y = \bar{m})$ is provable in S.

(1) For any $k \leq m$, the sentence $F(\bar{n}, \bar{k}) \supset \bar{m} \leq \bar{k}$ is provable, because if $k < m$, then $F(\bar{n}, \bar{k})$ is refutable, and if $k = m$, then $\bar{m} \leq \bar{k}$ is provable (by Ω_5). Then by Ω_4, the formula

$$z \leq \bar{m} \supset (F(\bar{n}, z) \supset \bar{m} \leq z)$$

is provable. Also

$$\bar{m} \leq z \supset (F(\bar{n}, z) \supset \bar{m} \leq z)$$

is, obviously, provable, and so by Ω_5, $F(\bar{n}, z) \supset \bar{m} \leq z$ is provable, and hence, $\forall z(F(\bar{n}, z) \supset \bar{m} \leq z)$ is provable. Also $F(\bar{n}, \bar{m})$ is provable and so

$$F(\bar{n}, \bar{m}) \wedge \forall z(F(\bar{n}, z) \supset \bar{m} \leq z)$$

is provable—i.e. $G(\bar{n}, \bar{m})$ is provable.

(2) The proof of (2) is obvious. For any $k \neq m$, $F(\bar{n}, \bar{k})$ is refutable. Hence $G(\bar{n}, \bar{k})$ is refutable (since $G(\bar{n}, \bar{k}) \supset F(\bar{n}, \bar{k})$ is, obviously, provable).

(3) To prove (3), we first show that the formula $G(\bar{n}, y) \supset y \leq \bar{m}$ is provable. Well,

$$G(\bar{n}, y) \supset \forall z(F(\bar{n}, z) \supset y \leq z)$$

is, obviously, provable. Hence

$$G(\bar{n}, y) \supset (F(\bar{n}, \bar{m}) \supset y \leq \bar{m})$$

is provable. But $F(\overline{n},\overline{m})$ is provable, and so by propositional logic, $G(\overline{n},y) \supset y \leq \overline{m}$ is provable.

Next we note that for any $k \leq m$, $G(\overline{n},\overline{k}) \supset \overline{k} = \overline{m}$ is provable because if $k < m$, then $G(\overline{n},\overline{k})$ is refutable (since $F(\overline{n},\overline{k})$ is refutable), and if $k = m$, then $\overline{k} = \overline{m}$ is provable. Then by Ω_4, the formula $y \leq \overline{m} \supset (G(\overline{n},y) \supset y = \overline{m})$ is provable. But $G(\overline{n},y) \supset y \leq \overline{m}$ is provable (as we have shown) and so $G(\overline{n},y) \supset y = \overline{m}$ is provable. Hence $\forall y(G(\overline{n},y) \supset y = \overline{m})$ is provable. This concludes the proof.

Since all formulas of Ω_4 and Ω_5 are provable in (R) and all recursive functions are weakly definable in (R), then by the above lemma, all recursive functions are strongly definable in (R), which establishes Theorem 3.

Proposition 3. *For any function $f(x_1,\ldots,x_n)$, if the relation*

$$f(x_1,\ldots,x_n) = x_{n+1}$$

is Σ_1, then the function $f(x_1,\ldots,x_n)$ is recursive.

Proof. Suppose the relation $f(x_1,\ldots,x_n) = x_{n+1}$ is Σ_1. Then the relation $f(x_1,\ldots,x_n) \neq x_{n+1}$ is also Σ_1, for it can be written as

$$\exists y(f(x_1,\ldots,x_n) = y \wedge y \neq x_{n+1}).$$

[This condition is obviously Σ and we know that all Σ-conditions are Σ_1.]

The Diagonal Function. We know that the diagonal function $d(x)$ is Σ_1. So by Proposition 3, it is recursive. Then by Theorem 3 (and the fact that any function strongly definable in a system is strongly definable in every extension of the system), we have

Proposition 4. *The diagonal function $d(x)$ is strongly definable in every extension of (R).*

§4. Fixed Points and Gödel Sentences.

For any expression X, we define $\overline{X}$ to be the numeral designating the Gödel number of X. Thus, for any formula $F(v_1)$ and any expression X, $F(\overline{X})$ is $F(\overline{x})$, where x is the Gödel number of X.

A sentence X is called a *fixed point* of a formula $F(v_1)$ (in a system $\mathcal{S}$) if the sentence $X \equiv F(\overline{X})$ is provable in $\mathcal{S}$.

Theorem 4. *If the diagonal function $d(x)$ is strongly definable in $\mathcal{S}$, then every formula $F(v_1)$ has a fixed point (in $\mathcal{S}$).*

Fixed Points and Gödel Sentences

Proof. Suppose the diagonal function $d(x)$ is strongly definable in S. Let $F(v_1)$ be any formula in v_1. By Theorem 2, there is then a formula $H(v_1)$ such that for any number n, $H(\overline{n}) \equiv F(\overline{d(n)})$ is provable in S. Then $H(\overline{h}) \equiv F(\overline{d(h)})$ is provable in S, where h is the Gödel number of $H(v_1)$. Therefore, $H[\overline{h}] \equiv F(\overline{d(h)})$ is provable in S (because the sentence $H[\overline{h}] \equiv H(\overline{h})$ is logically valid). But $d(h)$ is the Gödel number of $H[\overline{h}]$, and so $X \equiv F(\overline{X})$ is provable in S, where X is the sentence $H[\overline{h}]$.

Corollary 1. *If S is any extension of (R), every formula $F(v_1)$ has a fixed point in S.*

Gödel Sentences and Fixed Points. In Chapter 2, we defined a sentence X to be a *Gödel sentence* for a set A iff it is the case that X is true iff A contains the Gödel number of X. More generally, let us call X a Gödel sentence for A *with respect to a system S* if X is provable in S iff A contains the Gödel number of X. Thus, X is a Gödel sentence for A, in the sense of Chapter 2, if X is a Gödel sentence for A with respect to the complete theory $\mathcal{N}$.

Let us call a function $f(x)$ *acceptable* in S if for every set A representable in S, the set $f^{-1}(A)$ is also representable in S. The following theorem (a generalization of Theorem 1 of Chapter 2) is apparently of incomparable strength with Theorem 4 above; the hypothesis and the conclusion are both weaker (cf. discussion that follows).

Theorem 5. *If the diagonal function $d(x)$ is acceptable in S, then for every set A representable in S, there is a Gödel sentence for A.*

Proof. Exercise.[1]

Discussion. By statement (1) of the corollary of Theorem 2, if $f(x)$ is strongly definable in S, then $f(x)$ is acceptable in S. Therefore, the hypothesis of Theorem 4 is stronger than the hypothesis of Theorem 5. The conclusion of Theorem 4 is also stronger than the conclusion of Theorem 5; this can be seen as follows. To say that X is a Gödel sentence for the set represented by $F(v_1)$ in S is to say that X is provable in S iff $F(\overline{X})$ is provable in S (why?). To say that X is a *fixed point* of $F(v_1)$ is to say that the equivalence $X \equiv F(\overline{X})$ is actually provable in S. So a fixed point for $F(v_1)$ is automatically a Gödel sentence for the set represented in S by $F(v_1)$, but (in general) it is more.

---

[1] Cf. proof of Theorem 1, Chapter 2.

Exercise 5.[2] Suppose $F(x,y)$ is a formula that enumerates, not P^*, but the set P in $\mathcal{S}$, and X is any fixed point of the formula $\forall y \sim F(x,y)$. Show that if $\mathcal{S}$ is consistent, then X is not provable in $\mathcal{S}$, and if $\mathcal{S}$ is ω-consistent, then X is not refutable in $\mathcal{S}$. How can this be used to show that if P.A. is ω-consistent then P.A. is incomplete?

Exercise 6.[3] Suppose $\mathcal{S}$ is an extension of Ω_4 and Ω_5, and $F(x,y)$ enumerates P in $\mathcal{S}$ and $G(x,y)$ enumerates R in $\mathcal{S}$. Show that any fixed point of the formula

$$\forall y(F(x,y) \supset (\exists z \leq y)G(x,y))$$

is undecidable in $\mathcal{S}$ (assuming $\mathcal{S}$ is simply consistent).

Exercise 7. More generally, show that if $H(x)$ is a formula that represents in $\mathcal{S}$ some superset of R that is disjoint from P, then any fixed point of $H(x)$ is undecidable in $\mathcal{S}$. [We do not need to use Ω_4 and Ω_5 for this.]

Exercise 8. Prove the following theorem of Tarski: If the diagonal function $d(x)$ is strongly definable in $\mathcal{S}$ and $\mathcal{S}$ is consistent, then the set P is not definable in $\mathcal{S}$.

§5. Truth Predicates.

A formula $T(v_1)$ is called a *truth-predicate* for $\mathcal{S}$ if for every sentence X, the sentence $X \equiv T(\overline{X})$ is provable in $\mathcal{S}$.

The following two Tarski-type theorems appear to be incomparable in strength.

Theorem 6. *If $\mathcal{S}$ is correct (i.e., a subsystem of $\mathcal{N}$), then there is no truth predicate for $\mathcal{S}$.*

Theorem 7. *If $\mathcal{S}$ is any simply consistent system in which the diagonal function $d(x)$ is strongly definable, then there is no truth-predicate for $\mathcal{S}$.*

Proof of Theorem 6. Suppose there were a formula $T(v_1)$ such that for every sentence X, the sentence $X \equiv T(\overline{X})$ is provable in $\mathcal{S}$. If, furthermore, $\mathcal{S}$ is correct, then for any sentence X, the sentence $X \equiv T(\overline{X})$ must, therefore, be true. Hence X is true iff $T(\overline{X})$ is

---

[2] A Variant of Gödel's Proof

[3] A Variant of Rosser's Proof

true. This means that the formula $T(v_1)$ *expresses* the set of Gödel numbers of the true sentences, contrary to Tarski's theorem for $\mathcal{L}_A$.

Proof of Theorem 7. Suppose $d(x)$ is strongly definable in $\mathcal{S}$, and $T(v_1)$ is a truth-predicate for $\mathcal{S}$. By Theorem 4, there must be a sentence X (a fixed-point for $\sim T(v_1)$) such that $X \equiv \sim T(\overline{X})$ is provable. But also $X \equiv T(\overline{X})$ is provable (since $T(v_1)$ is a truth-predicate). Hence it follows that $T(\overline{X}) \equiv \sim T(\overline{X})$ is provable in $\mathcal{S}$, which means that $\mathcal{S}$ is inconsistent.

Chapter IX

The Unprovability of Consistency

Gödel's second incompleteness theorem, roughly stated, is that if Peano Arithmetic is consistent, then it cannot prove its own consistency.[1] The theorem has been generalized and abstracted in various ways and this has led to the notion of a *provability predicate*, which plays a fundamental role in much modern metamathematical research. To this notion we now turn.

§1. **Provability Predicates.** A formula $P(v_1)$ is called a *provability predicate* for S if for all sentences X and Y the following three conditions hold:

P_1: If X is provable in S, then so is $P(\overline{X})$.
P_2: $P(\overline{X \supset Y}) \supset (P(\overline{X}) \supset P(\overline{Y}))$ is provable in S.
P_3: $P(\overline{X}) \supset P(\overline{P(\overline{X})})$ is provable in S.

Suppose now $P(v_1)$ is a Σ_1-formula that expresses the set P of the system P.A. Under the assumption of ω-consistency, $P(v_1)$ represents P in P.A. Under the weaker assumption of simple consistency, all that follows is that $P(v_1)$ represents some *superset* of P, but that is enough to imply that if X is provable in P.A., then so is $P(\overline{X})$. Therefore property P_1 holds. As for property P_2, the sentence $P(\overline{X \supset Y}) \supset (P(\overline{X}) \supset P(\overline{Y}))$ is obviously true (its truth is equivalent to the proposition that if $X \supset Y$ and X are both provable in P.A., then Y is provable in P.A., which, of course, is the case since modus ponens is an inference rule of P.A.). It is not very difficult to formalize this argument and show that the above sentence is not only true, but provable in P.A.

---

[1] A precise formulation of this theorem will be given in this chapter.

As for property P_3, the sentence $P(\overline{X}) \supset P(\overline{P(\overline{X})})$ is, of course, true (its truth is equivalent to the proposition that if X is provable, then so is $P(\overline{X})$ because $P(\overline{X})$ is true iff X is provable; $P(\overline{P(\overline{X})})$ is true iff $P(\overline{X})$ is provable). So the *truth* of the sentence reduces to property P_1. The sentence is not only true but even provable in P.A., but the proof of this fact is extremely elaborate and goes beyond the scope of this volume. The fact is a special case of the well-known fact that for every Σ_1-sentence Y, the sentence $Y \supset P(\overline{Y})$ is provable in P.A. A sketch of the proof of this can be found in Ch. 2 of Boolos [1979]. A detailed treatment for a system akin to P.A. can be found in Hilbert-Bernays [1934–39] and a helpful discussion can be found in Shoenfield [1967].

Until further notice, it will be assumed that $P(v_1)$ is a provability predicate for $\mathcal{S}$. Provability predicates enjoy the following three properties (for any sentences X, Y and Z).

P_4: If $X \supset Y$ is provable in $\mathcal{S}$, then so is $P(\overline{X}) \supset P(\overline{Y})$.
P_5: If $X \supset (Y \supset Z)$ is provable in $\mathcal{S}$, then so is

$$P(\overline{X}) \supset (P(\overline{Y}) \supset P(\overline{Z})).$$

P_6: If $X \supset (P(\overline{X}) \supset Y)$ is provable in $\mathcal{S}$, then so is

$$P(\overline{X}) \supset P(\overline{Y}).$$

Proof.

P_4: Suppose $X \supset Y$ is provable (in $\mathcal{S}$). Then so is $P(\overline{X \supset Y})$ (by property P_1). Then $P(\overline{X}) \supset P(\overline{Y})$ is provable (using property P_2 and modus ponens).

P_5: Suppose $X \supset (Y \supset Z)$ is provable. Then so is

$$P(\overline{X}) \supset P(\overline{Y \supset Z})$$

(by P_4). Also $P(\overline{Y \supset Z}) \supset (P(\overline{Y}) \supset P(\overline{Z}))$ is provable (by P_4). Hence by propositional logic, $P(\overline{X}) \supset (P(\overline{Y}) \supset P(\overline{Z}))$ is provable.

P_6: Suppose $X \supset (P(\overline{X}) \supset Y)$ is provable. Then

$$P(\overline{X}) \supset (P(\overline{P(\overline{X})}) \supset P(\overline{Y}))$$

is provable (by P_5). Also $P(\overline{X}) \supset P(\overline{P(\overline{X})})$ is provable. From the last two facts, it follows, by propositional logic, that

$$P(\overline{X}) \supset P(\overline{Y})$$

is provable in $\mathcal{S}$.

Properties P_1 and P_6 will play the key roles in that which follows.

We shall call $\mathcal{S}$ *diagonalizable* if every formula $F(v_1)$ has a fixed point (with respect to $\mathcal{S}$). Since P.A. is an extension of (R), then P.A. is diagonalizable by the corollary of Theorem 4 of the last chapter. We are now interested in provability predicates for diagonalizable systems.

§2. The Unprovability of Consistency.

We continue to assume that $P(v_1)$ is a provablity predicate for $\mathcal{S}$.

Theorem 1. *If G is a fixed point of the formula $\sim P(v_1)$ and $\mathcal{S}$ is consistent, then G is not provable in $\mathcal{S}$.*

Proof. We are given that $G \equiv \;\sim P(\overline{G})$ is provable in $\mathcal{S}$. Now suppose G were provable in $\mathcal{S}$. Then $\sim P(\overline{G})$ and $P(\overline{G})$ would be provable in $\mathcal{S}$ (by property P_1). Hence $\mathcal{S}$ would be inconsistent. So if $\mathcal{S}$ is consistent, then G is not provable in $\mathcal{S}$.

Consistency Sentences. We let "f" stand for any logical falsehood (such as any sentence of the form $X \wedge \sim X$)—or for that matter, any sentence refutable in $\mathcal{S}$. [A common choice for f in the system P.A. is the sentence $(\overline{0} = \overline{1})$.] We let consis be the sentence $\sim P(\overline{f})$.

If $P(v_1)$ is a "correct" provability predicate for $\mathcal{S}$ (i.e., if $P(v_1)$ expresses the set P), then the sentence consis is true iff f is not provable in $\mathcal{S}$—in other words, iff $\mathcal{S}$ is consistent. So the sentence consis is an arithmetic sentence which can be said to "express" the consistency of $\mathcal{S}$. The next two theorems, however, do not require the assumption that $P(v_1)$ is a *correct* provability predicate for $\mathcal{S}$ but only that $P(v_1)$ is a provability predicate for $\mathcal{S}$.

The next theorem is a key lemma for the proof of Gödel's Second Incompleteness Theorem.

Theorem 2. *If G is a fixed point of the formula $\sim P(v_1)$, then the sentence consis $\supset G$ is provable in $\mathcal{S}$.*

Proof. We are given that the sentence $G \equiv \;\sim P(\overline{G})$ is provable in $\mathcal{S}$. Since f is a refutable sentence of $\mathcal{S}$, then $\sim P(\overline{G}) \equiv (P(\overline{G}) \supset f)$ is provable in $\mathcal{S}$ and, therefore, $G \equiv (P(\overline{G}) \supset f)$ is provable in $\mathcal{S}$, and so $G \supset (P(\overline{G}) \supset f)$ is provable. Then by property P_6, the sentence $P(\overline{G}) \supset P(\overline{f})$ is provable. Hence $\sim P(\overline{f}) \supset\;\sim P(\overline{G})$ is provable, and since $G \equiv \sim P(\overline{G})$ is provable, then $\sim P(\overline{f}) \supset G$ is provable. Thus, consis $\supset G$ is provable in S.

From Theorems 1 and 2 we have:

Theorem 3—An Abstract Form of Gödel's Second Theorem. Suppose S is diagonalizable. Then if S is consistent, then the sentence consis is not provable in S.

Proof. Suppose S is diagonalizable. Then there is a sentence G such that $G \equiv \sim P(\overline{G})$ is provable in S and by Theorem 2, the sentence consis $\supset G$ is provable in S. Now suppose consis were provable in S. Then G would be provable in S and S would be inconsistent by Theorem 1 (since G is a fixed point of $\sim P(v_1)$). Therefore, if S is consistent, then the sentence consis is not provable in S.

Discussion. For the case when S is the system P.A. and $P(v_1)$ is a Σ_1-formula expressing the set P, the sentence consis is a true sentence (assuming P.A. to be consistent), but it is not provable in P.A. This result has been paraphrased, "If arithmetic is consistent, then it cannot prove its own consistency." Unfortunately there has been a good deal of popular nonsense written about this by writers who, obviously, do not understand what the matter is all about. We have seen such irresponsible statements as, "By Gödel's second theorem, we can never know whether or not arithmetic is consistent." Rubbish! To see how silly this is, suppose it had turned out that the sentence consis were provable in P.A.—or, to be more realistic, suppose we consider a system that can prove its own consistency. Would that be any grounds for trusting the consistency of the system? Of course not! If the system were inconsistent, then it could prove every sentence—including the statement of its own consistency! To trust the consistency of a system on the grounds that it can prove its own consistency is as foolish as trusting a person's veracity on the grounds that he claims that he never lies. No, the fact that P.A., if consistent, cannot prove its own consistency—this fact does not constitute the slightest rational grounds for doubting the consistency of P.A.

§3. Henkin Sentences and Löb's Theorem.

Leon Henkin [1952] raised the following famous questions about the system P.A.: Since the system is diagonalizable, there is a fixed point for the formula $P(v_1)$—a sentence H such that $H \equiv P(\overline{H})$ is provable in P.A. Gödel's sentence G—a fixed point of $\sim P(v_1)$—is true iff it is *not* provable in P.A.; Henkin's sentence H is true iff it *is* provable in P.A. Thus, H is either true and provable in P.A. or false and not provable in P.A. Is there any way to tell which? This problem was answered by Löb [1955] who showed that the provability of $P(\overline{H}) \supset H$ (let alone that of $P(\overline{H}) \equiv H$) is enough to guarantee the provability of H. Here is Löb's theorem (we continue to assume that $P(v_1)$ is a provability predicate for $\mathcal{S}$).

Theorem 4—Löb's Theorem. *Suppose $\mathcal{S}$ is diagonalizable and $P(v_1)$ is a provability predicate for $\mathcal{S}$. Then for any sentence Y, if the sentence $P(\overline{Y}) \supset Y$ is provable in $\mathcal{S}$, then so is Y.*

Proof. Assume hypothesis. Now suppose $P(\overline{Y}) \supset Y$ is provable in $\mathcal{S}$. Since $\mathcal{S}$ is diagonalizable, there is a fixed point X for the formula $P(v_1) \supset Y$—thus, $X \equiv (P(\overline{X}) \supset Y)$ is provable. Therefore, $X \supset (P(\overline{X}) \supset Y)$ is provable. So by property P_6, the sentence $P(\overline{X}) \supset P(\overline{Y})$ is provable. Then, since the sentence $P(\overline{Y}) \supset Y$ is assumed provable, the sentence $P(\overline{X}) \supset Y$ is provable. But $X \equiv (P(\overline{X}) \supset Y)$ is provable and since $P(\overline{X}) \supset Y$ is provable, so is X. Therefore, $P(\overline{X})$ is provable (by property P_1), and since $P(\overline{X}) \supset Y$ is provable, so is Y.

Georg Kreisel has pointed out that Gödel's second theorem can be obtained as an easy corollary of Löb's Theorem. Suppose consis is provable in $\mathcal{S}$. Consis is the sentence $\sim P(\overline{f})$, and so the sentence $P(\overline{f}) \supset f$ is provable in $\mathcal{S}$. Then by Löb's theorem, f is provable in $\mathcal{S}$, which means that $\mathcal{S}$ is inconsistent!

It has also been observed by Saul Kripke that Löb's Theorem can be obtained as a corollary of Gödel's second theorem (applied to $\mathcal{S}$ and extensions of $\mathcal{S}$).[2]

In the last decade or so, the whole subject of provability predicates for diagonalizable systems has been tied up with the study of modal logic. The union has been most fruitful! An excellent account of all this can be found in Boolos: *The Unprovability of Consistency*, which we earnestly recommend to the reader as a perfect follow-up to this chapter.

Exercise 1. In this and the following exercises, $P(v_1)$ is assumed to be a provability predicate for $\mathcal{S}$.

Show that if $X \supset (P(\overline{X}) \supset Y)$ is provable in $\mathcal{S}$, then

$$(P(\overline{Y}) \supset Y) \supset (P(\overline{X}) \supset Y)$$

---

[2] cf. Ch. 16 of Boolos and Jeffrey for an interesting discussion of this point

is provable in S.

Exercise 2. Show that if $X \equiv (P(\overline{X}) \supset Y)$ is provable (in S), then so is $(P(\overline{Y}) \supset Y) \supset X$.

Exercise 3. Show that Th. 2 is a special case of Ex. 2.

Exercise 4. Show that for any sentence X, the sentence

$$\sim P(\overline{X}) \supset \text{consis}$$

is provable in S.

Exercise 5. Show that under the hypothesis of Theorem 2, the sentence consis $\equiv G$ is provable in S.

Exercise 6. Show that if $X \equiv Y$ is provable in S, then

$$P(\overline{X}) \equiv P(\overline{Y})$$

is provable in S.

Exercise 7. Show that if S is diagonalizable, then the sentence consis $\supset \sim P(\overline{\text{consis}})$ is provable in S.

Exercise 8. Show that if S is consistent and diagonalizable, then for no sentence X is it the case that $\sim P(\overline{X})$ is provable in S.

Chapter X

Some General Remarks on Provability and Truth

We have given three different incompleteness proofs of Peano Arithmetic—the first used Tarski's truth-set, the second (Gödel's original proof) was based on the assumption of ω-consistency, and the third (Rosser's proof) was based on the assumption of simple consistency. The three proofs yield different generalizations—namely

1. Every axiomatizable subsystem of $\mathcal{N}$ is incomplete.
2. Every axiomatizable ω-consistent system in which all true Σ_0-sentences are provable is incomplete.
3. Every axiomatizable simply consistent extension of (R) is incomplete.

The first of the three proofs is by far the simplest and we are surprised that it has not appeared in more textbooks. Of course, it can be criticized on the grounds that it is not formalizable in arithmetic (since the truth set is not expressible in arithmetic), but this should be taken with some reservations in light of Askanas' theorem, which we will discuss a bit later.

Induction and the ω-Rule. It is not too surprising that Peano Arithmetic is incomplete because the scheme of mathematical induction does not really express the full force of mathematical induction. The true principle of mathematical induction is that for *any* set A of natural numbers, if A contains 0 and A is closed under the successor function (such a set A is sometimes called an *inductive* set), then A contains all natural numbers. Now, there are non-denumerably many sets of natural numbers but only denumerably many formulas in the language L_A and, hence, there are only denumerably many *expressible* sets of L_A. Therefore, the formal axiom scheme of induction for P.A. guarantees only that for every *expressible* set A, if A is inductive, then A contains all natural numbers.

To express the principle of mathematical induction fully, we need *second order* arithmetic in which we take set and relational variables and quantify over sets and relations of natural numbers. Then we can wholly express true mathematical induction using just the one formula
$$\forall A((A\bar{0} \wedge \forall v_1(Av_1 \supset Av_1')) \supset \forall v_1 Av_1).$$
The reader might now wonder whether the system consisting of the Peano Axioms couched in second-order logic is complete. The answer is that it is not because although mathematical induction is fully expressible in second-order arithmetic, the trouble is that the underlying logic (second-order logic) is not axiomatizable—i.e., the set of Gödel numbers of the logically valid second-order formulas is not Σ_1.[1]

Returning to first-order arithmetic, suppose we added to P.A. the following inference rule (known as the ω-rule; it is sometimes called *Tarski's* rule and sometimes *Carnap's* rule): For any formula $F(v_1)$, one may infer $\forall v_1 F(v_1)$ from the infinitely many premises $F(\bar{0}), F(\bar{1}), \ldots, F(\bar{n}), \ldots$. Let us call this system P.A.$^+$ Gödel's sentence G is provable in P.A.$^+$ (since it is a universal sentence all of whose instances are provable in P.A.). Indeed, it is not hard to see that *all* true sentences are provable in P.A.$^+$ and P.A.$^+$ is complete. Why not, then, use P.A.$^+$ instead of P.A. as a working axiom system? The answer is that there is no way it can be *used* by a finite being like a human or a computer. Proofs in P.A.$^+$ are (sometimes) of transfinite length.

We could add to P.A. a weaker rule to the effect that for any formula $F(v_1)$, if it is *provable* in P.A. that every instance of $F(v_1)$ is provable in P.A., then $\forall v_1 F(v_1)$ may be inferred. In this system, Gödel's sentence G is indeed provable, but the system is axiomatizable, hence another sentence G_1 can be found which is not provable in that system.

[To state this rule more accurately, one can associate with each formula $F(v_1)$ a Σ_1-formula $F^*(v_1)$ that expresses the set of all n such that $F(\bar{n})$ is provable in P.A. Then one can add the axiom scheme $\forall v_1 F^*(v_1) \supset P(\overline{\forall v_1 F(v_1)})$.]

---

[1] See Boolos and Jeffrey for a further discussion of this point.

There is simply no way to get around the fact that $\mathcal{N}$ is not axiomatizable, hence every axiomatizable subsystem of $\mathcal{N}$ is incomplete.

Some Remarks on Arithmetic Truth.

1. For each n, let T_n be the set of Gödel numbers of all true sentences of degree n or less. The set T (of Gödel numbers of all true sentences) is then the union of all the sets $T_0, T_1, \ldots, T_n, \ldots$. We know that the set T is not arithmetic, yet for each individual number n, the set T_n *is* arithmetic. [Hint: For each n, let F_n be the set of Gödel numbers of the false sentences of degree n or less. It is easy to construct formulas $T_0(v_1)$ and $F_0(v_1)$, which express the sets T_0 and F_0 respectively. Then, for each n, given formulas $T_n(v_1)$ and $F_n(v_1)$ respectively expressing the sets T_n and F_n, it is not difficult to find formulas $T_{n+1}(v_1)$ and $F_{n+1}(v_1)$ that respectively express the sets T_{n+1} and F_{n+1}. Details can be found in Chapter 19 of Boolos and Jeffrey.]

2. Let us add one predicate variable M of degree 1 to the language of P.A. [Thus, the atomic formulas are those of L_A together with any formula Mt, where t is any term.] Let $\Phi(M)$ be any closed formula with M as a free predicate variable (all individual variables v_i are bound). The sentence $\Phi(M)$ is neither true nor false but becomes either true or false when "M" is interpreted as the name of some set of natural numbers.

    It is not difficult to construct a formula $J(M)$ such that if "M" is interpreted as being the name of the set T, then $J(M)$ is true, and under any other interpretation of "M", the sentence $J(M)$ is false. [A proof of this can be found in Ch. 19 of Boolos and Jeffrey, but the reader should try it as an exercise.] It follows from this that the set T, though not expressible in first-order arithmetic (Tarski's theorem), is expressible in second-order arithmetic—expressed, in fact, by the formula $\exists M(J(M) \wedge Mv_1)$ (or alternatively by $\forall M(J(M) \supset Mv_1)$).

3. *Askanas' Theorem.* Suppose we now take any formula F in one free variable v_1 and for each variable v_i, we replace every occurrence of Mv_1 in $J(M)$ by $F(v_1)$; we then obtain an arithmetic sentence—call it "$J(F)$"—that expresses the proposition that $F(v_1)$ expresses the set T. By Tarski's theorem, no formula $F(v_1)$ expresses the set T. Hence for every formula $F(v_1)$, the sentence $J(F)$ is false, and the sentence $\sim J(F)$ is true. Askanas' theorem [1975] is that for every formula F, the sentence $\sim J(F)$ is not only true, but provable in Peano Arith-

metic. Thus, in a sense, Tarski's theorem is formalizable in first-order arithmetic even though T is not arithmetic. [Roughly speaking, Askanas' Theorem bears much the same relation to Tarski's theorem as Gödel's second theorem bears to Gödel's first incompleteness theorem.]

It was indicated in an Exercise in Ch. III how to obtain an arithmetic formula $P_s(v_1)$ that expresses the set of Gödel numbers of the provable *sentences* of P.A. (or rather, this was done for P.E., but the modifications for P.A. are obvious). Then by Askanas' theorem, the sentence $\sim J(P_s)$ is provable in P.A. This sentence is true iff the set of provable sentences of P.A. differs from the set of true sentences. Hence, under the assumption that P.A. is correct, it follows that some true sentence is not provable in P.A. This constitutes a slight variant of the Gödel-Tarski incompleteness proof, but has the advantage of showing that $\sim J(P_s)$ is not only true, but actually provable in P.A.

Exercise. [For readers with some familiarity with second-order logic] Many relations which can be shown to be expressible in first-order arithmetic can be far more easily shown to be expressible in second-order arithmetic. For example, we went through much labor in showing that the exponential relation $x^y = z$ is (first-order) expressible from plus and times. [We needed the finite sequence lemma, or a β-function to do this.] In second-order arithmetic, it can be done very easily in the following manner.

Consider second-order arithmetic in which we have variables for functions (as well as for sets and relations). Let "f" be a functional variable of two arguments and let $E(f)$ be the sentence

$$\forall v_1 f(v_1, \overline{0}) = \overline{1} \wedge \forall v_1 \forall v_2 (f(v_1, v_2') = f(v_1, v_2) \cdot v_1).$$

1. Show that if f is interpreted as the exponential function x^y, then the sentence $E(f)$ is true, but if f is interpreted as any other function, the sentence is false.
2. Show that the second-order formula $\exists f(E(f) \wedge f(v_1, v_2) = v_3)$ expresses the relation $x^y = z$. Show the same for the formula $\forall f(E(f) \supset f(v_1, v_2) = v_3)$.

Chapter XI

Self-Referential Systems

This chapter is largely a review of the essential ideas behind the proofs of Gödel, Rosser and Löb—only presented in a more abstract setting. We believe that it will tie up these ideas in a helpful and instructive manner.

We shall first present these ideas in the form of logic puzzles (much in the manner of Smullyan [1987]). Then we shall state the results more generally in terms of abstract systems that we call *provability systems*. These are closely related to certain axiom systems of modal logic, which we briefly discuss at the end of the chapter.

I. *Logicians Who Reason About Themselves*

In the puzzles to which we now turn, *belief* will play the rôle of provability. Instead of considering a mathematical system and the sentences provable in it, we consider a logician (sometimes call a *reasoner*) and the propositions believed by the reasoner. Apart from the heuristic value, these "epistemic" incompleteness theorems appear to be of some interest to those working in artificial intelligence.

§1. An Analogue of the Tarski-Gödel Theorem. We shall pay a visit to the Island of *Knights and Knaves*, in which knights make only true statements and knaves make only false ones. Each inhabitant is either a knight or a knave. No inhabitant can claim that he is not a knight (since a knight would never make such a false claim and a knave would never make such a true claim).

A logician visits this island one day and meets a native. All we are told about the logician is that he is completely accurate in his beliefs—he never believes anything false. The native then makes a certain statement X. It then follows that the logician can never believe that the native is a knight nor can he ever believe that the native is a knave.

Problem 1. What statement X would accomplish this?

A Solution. One solution is that the native says, "You will never believe that I'm a knight." If the native were a knave, then his statement would be false, which would mean that the logician *would* believe that the native is a knight, contrary to the assumption that the logician never believes anything false. Therefore, the native must be a knight. It, then, further follows that the native's statement was true and, hence, the logician can never believe that the native is a knight. Then, since the native really is a knight and the logician believes only true statements, he also will never believe that the native is a knave. And so the logician must remain forever undecided as to whether the native is a knight or a knave.

Discussion and Terminology. The function of the Knight-Knave island was to get an "easy" fixed-point (the native can use the indexical term "I"). In all the problems that follow, there will be one native and one logician. We shall let k be the proposition that the native is a knight. Now, whenever the native asserts a proposition p, the proposition $k \equiv p$ must be true (if the native is a knight, then p must be true, and if p is true, then the native must be a knight). For any proposition p, we let Bp be the proposition that the logician will sooner or later believe p. Now, in the above problem, the native has asserted $\sim Bk$ (you will never believe that I'm a knight) and so the proposition $k \equiv \sim Bk$ is true. Actually, for *any* proposition q, if $q \equiv \sim Bq$ is true and the logician never believes false propositions, then he can never believe q nor can he ever believe $\sim q$. Thus, q must be true, but the logician can never believe q.

§2. Normal and Stable Reasoners of Type 1.

In the problem just considered, it was not necessary that the reasoner actually heard the statement made by the native. Indeed, even if the reasoner—call him Robert—were thousands of miles from the island and a native of the island said, "Robert will never believe that I'm a knight," then Robert, if always accurate in his beliefs, can never believe that the native is a knight nor can he ever believe that the native is a knave. In the problems that follow, it will be necessary to assume that the reasoner *is* on the island and hears the statement made by the native. Moreover, it is to be assumed that the reasoner *believes* that knights make only true statements, knaves make only false ones and that every inhabitant is either a knight or a knave. And so when the native asserts a proposition p to the reasoner, the reasoner will *believe* the proposition $k \equiv p$ when k is the proposition that the native is a knight.

As a matter of fact, in all the problems that follow, it is no longer necessary to assume that knights really make true statements and that knaves make false ones; all that it is necessary to assume is that the reasoner *believes* that this is the case. Indeed, the notion of *truth* will no longer be relevant; all that matters is what the reasoner believes. [We are getting away from the Tarski-Gödel argument and veering towards the Gödel and Rosser arguments.] And so, to repeat a vital point, when the native asserts a proposition p, the reasoner will believe $k \equiv p$.

Reasoners of Type 1. We shall say that the reasoner is of *type 1* if he has a thorough knowledge of propositional logic—i.e. the set of all propositions that he ever believes contains all tautologies and is closed under modus ponens (if he ever believes p and ever believes $p \supset q$, then he will believe q). Of course, the assumption that the reasoner is of type 1 is highly idealized since there are infinitely many tautologies, and so we must assume something like immortality on the reasoner's part (but little things like that don't bother us in mathematics).

We shall also credit the reasoner with another power (which, though not necessary, will make some of our later arguments shorter and more transparent)—namely that the reasoner can do natural deduction—i.e., if by assuming p as a premiss, the reasoner can derive q as a conclusion, then the reasoner will believe $p \supset q$. [By a well-known result called the *deduction theorem*, this extra power will not enlarge the class of the reasoner's beliefs.]

Normality and Stability. We shall call the reasoner *normal* if whenever he believes a proposition p, he believes that he believes p (thus if he believes p, then he believes Bp). We shall call him *stable* if the converse holds—i.e., if he believes Bp (if he believes that he believes p), then he really does believe p. We will call the reasoner *unstable* if he is not stable—i.e., if there is at least one proposition p such that the reasoner believes that he believes p, but doesn't actually believe p. [We remark that if a reasoner is unstable, he believes

I. Logicians Who Reason About Themselves

at least one false proposition, and so an accurate reasoner is automatically stable. But stability is a weaker condition than accuracy and does not involve the notion of truth. As we will see a bit later, instablity is closely related to ω-inconsistency.]

We shall call the reasoner *inconsistent* if he believes some proposition p and also believes $\sim p$. For a reasoner of type 1, this is equivalent to the condition that he believes every proposition (sooner or later) since $p \supset (\sim p \supset q)$ is a tautology. Also, for any tautologically contradictory proposition f, a reasoner of type 1 is inconsistent if and only if he ever believes f (since for every q, the proposition $f \supset q$ is a tautology). We call the reasoner *consistent* if he is not inconsistent.

We might remark that a stable reasoner is not necessarily consistent (in fact, an inconsistent reasoner of type 1 is automatically stable since he will believe everything!) Nor is a consistent reasoner necessarily stable.

Now for the problem.

Problem 2. A normal reasoner of type 1 visits the Island of Knights and Knaves and meets a native who tells him, "You will never believe that I'm a knight."

Prove that if the reasoner is both consistent and stable, then he must remain forever undecided as to whether the native is a knight or a knave. More specifically prove

1. If the reasoner ever believes that the native is a knight, then he will be inconsistent.
2. If the reasoner ever believes that the native is a knave, then he will be either inconsistent or unstable. [Solution is given following Problem 3.]

ω-Consistent Reasoners. In the two problems above, it made no difference in what order the reasoner believed various propositions. In this problem and the next, the order will be relevant.

The reasoner arrives on the Island of Knights and Knaves on some day which we call the 0th day and he believes various propositions on various days after his arrival. For any proposition p and any natural number n, we let $B_n p$ be the proposition that the reasoner believes p on the nth day (after his arrival). We continue to let Bp be the proposition that the reasoner believes p on some day or other. [Thus, Bp is the proposition $\exists n B_n p$.] We will call the reasoner ω-inconsistent if there is at least one proposition p such that the reasoner (sooner or later) believes Bp, yet for each natural number n, he (sooner or later) believes $\sim B_n p$. For a reasoner of type 1, if he is (simply) inconsistent, then he will sooner or later believe every proposition and will also be ω-inconsistent. Thus an ω-consistent reasoner of type 1 is also simply consistent.

For the present problem, we are given that the reasoner satisfies the following three conditions (n is any natural number; p is any proposition).

C_1: If he believes p on the nth day (after his arrival), then he will believe $B_n p$.
C_2: If he fails to believe p on the nth day, then he will believe $\sim B_n p$.
C_3: If he ever believes $B_n p$, then he will believe Bp.

The idea behind C_1 and C_2 is that the reasoner has perfect memory for what propositions he has or has not believed on past days. As to C_3, this tells us that the reasoner knows at least a tiny bit of first-order logic (he can pass from believing $B_n p$, for a particular n, to believing $\exists n B_n p$).

Problem 2A—After Gödel. A reasoner of type 1 satisfying conditions C_1, C_2 and C_3 arrives on the island on the 0th day and meets a native who tells him, "You will never believe that I am a knight." Prove

1. If the reasoner ever believes that the native is a knight, he will be inconsistent.
2. If the reasoner ever believes that the native is a knave, then he will be ω-inconsistent. [Solution is given following the solution of Problem 2.]

§3. Rosser Type Reasoners.

We continue to assume that the reasoner believes various propositions on various days.

For any propositions p and q, we will say that the reasoner believes p *before* he believes q—in symbols, $Bp < Bq$—if for some n, he believes p on the nth day, but has not yet believed q. We shall understand that if the reasoner ever believes p and never believes q, then the reasoner does believe p before he believes q (because on the first day that he believes p, he has not believed q on either that day or any earlier day). We note that $Bp < Bq$ and $Bq < Bp$ cannot both be true.

By a *Rosser-type* reasoner, we shall mean a reasoner of type 1 such that for any propositions p and q, if he believes p before he believes

q, then he will believe $Bp < Bq$ and $\sim (Bq < Bp)$. [Again the idea is that on any day, the reasoner has perfect memory for what he has and has not believed on that day and all earlier days.]

Problem 3—After Rosser. A Rosser-type reasoner visits the Island of Knights and Knaves and is told by a native, "You will never believe I'm a knight before you believe I'm a knave." [Symbolically, the native is asserting $\sim (Bk < B\sim k)$.]

Prove that if the reasoner is simply consistent, then he must remain forever undecided as to whether the native is a knight or a knave (if he should believe either one, he will be inconsistent).

Solution to Problem 2. Since the native has asserted $\sim Bk$, then the reasoner believes $k \equiv \sim Bk$.

1. Suppose he ever believes k. Then, since he believes $k \equiv \sim Bk$ and is of type 1, he will believe $\sim Bk$. But since he believes k and is normal, he will also believe Bk. Hence he will be inconsistent.
2. Suppose he ever believes $\sim k$. Then he will believe Bk (since he believes $k \equiv \sim Bk$ and, hence, $\sim k \equiv Bk$, since he is of type 1). If he is stable, then he will then believe k and, hence, he will be inconsistent. And so he is either unstable or inconsistent.

Solution to Problem 2A. The reasoner believes $k \equiv \sim Bk$.

1. Suppose he ever believes k. Then for some n, he believes k on the nth day. Hence by C_1, he will believe $B_n k$ and by C_3, he will believe Bk. But since he believes both k and $k \equiv \sim Bk$, he will believe $\sim Bk$ and thus be inconsistent.
2. Suppose he ever believes $\sim k$. Then he will believe Bk (since he is of type 1 and believe $k \equiv \sim Bk$). If he is simply consistent, then he will never believe k, and so for each n, he fails to believe k on the nth day. By C_2, for each n he will believe $\sim B_n k$. Yet he believes Bk, and so he is ω-inconsistent. Thus if he is simply consistent, then he is ω-inconsistent, and if he is not simply consistent, then he is certainly ω-inconsistent. Thus he is ω-inconsistent.

Remark. Actually the solution is a corollary of the result of Problem 2, because conditions C_1 and C_3 jointly imply that the reasoner is normal. Condition C_2 implies that if the reasoner is ω-consistent, then he must be stable. We leave the verification of these two facts to the reader.

Solution to Problem 3. For any proposition p, let $Q(p)$ be the proposition $Bp < B\sim p$. It follows from the definition of a Rosser-type reasoner that if he ever believes p, then he will believe $Q(p)$, and if he ever believes $\sim p$, then he will believe $\sim Q(p)$. The reason is this. Suppose he believes p. If he is inconsistent, then he will believe everything, including $Q(p)$. If he is consistent, then he will never believe $\sim p$. Hence he will believe p before he believes $\sim p$ and he will believe $Bp < B\sim p$, which is the proposition $Q(p)$. Similarly, if he ever believes $\sim p$, then he will believe $\sim Q(p)$ (he will also believe $Q(\sim p)$, but this won't help us in our solution).

Now, the native has asserted $\sim Q(k)$, and so the reasoner believes $k \equiv \sim Q(k)$. If he ever believes k, then he will then believe $\sim Q(k)$ and he will also believe $Q(k)$ (as we showed above). Hence, he will be inconsistent. If he ever believes $\sim k$, then he will believe $Q(k)$ and also $\sim Q(k)$ (as shown above), and he will again be inconsistent.

§4. The Consistency Problem.

In this problem and the next, we no longer need to consider the order in which the reasoner believes various propositions, but we will have to postulate some "introspective" properties of the reasoner (which are the analogues of a formula $P(v_1)$ being a provability predicate for a first-order system S).

By an *advanced reasoner*—or a reasoner of *type 4*—we shall mean a reasoner of type 1 who satisfies the following three conditions (where p and q are any propositions).

P_1: If he believes p, then he believes Bp (he is normal).
P_2: He believes $B(p \supset q) \supset (Bp \supset Bq)$.
P_3: He believes $Bp \supset BBp$.

We will paraphrase P_2 by saying that the reasoner knows that his beliefs are closed under modus ponens. [He believes, "If I should ever believe p and believe that p implies q, then I will believe q."] We shall paraphrase P_3 by saying that the reasoner knows that he is normal. [For any proposition p, the reasoner believes, "If I ever believe p, then I will believe that I believe p."] Of course, an advanced reasoner believes BX for any tautalogy X (since he is of type 1 and normal). And so, roughly speaking, an advanced reasoner is a normal reasoner of type 1 who knows that he is a normal reasoner of type 1.

We shall say that a reasoner believes that he is consistent if for every proposition p, he believes that he doesn't believe both p and

I. Logicians Who Reason About Themselves

$\sim p$. For a reasoner of type 1, this is equivalent to his believing that he will never believe f (where f can be any fixed tautologically false proposition). Thus, a reasoner of type 1 believes that he is consistent iff he believes $\sim Bf$.

We note that if an advanced reasoner believes $p \supset q$, then he will believe $Bp \supset Bq$ (because, being normal, he will believe

$$B(p \supset q),$$

and, hence, by P_2 and the use of modus ponens, he will believe $Bp \supset Bq$). Therefore if he believes $p \equiv q$, then he will believe both $Bp \supset Bq$ and $Bq \supset Bp$. Thus, if he visits the Island of Knights and Knaves and if a native asserts a proposition q, then he will not only believe $k \equiv q$ (as would any reasoner of type 1) but will also believe $Bp \supset Bq$ (as well as $Bq \supset Bp$). Thus he will believe, "If I ever believe that the native is a knight, then I will believe what he said" (and he will also believe the converse).

Problem 4—After Gödel's Second Theorem. An advanced reasoner visits the Island of Knights and Knaves and meets a native who tells him, "You will never believe I'm a knight." Prove that if the reasoner should believe that he is consistent, then he will become inconsistent. Stated otherwise, if the reasoner is (and remains) consistent, he can never know it.

Solution. We could give the solution in the formal manner of Chapter IX, but we find it far more perspicuous to use the fact that the reasoner can do natural deduction. It should be understood that the reasoner is so "programmed" that any argument he *can* use, he sooner or later *will* use.

And so, suppose the reasoner assumes his own consistency. Then he will sooner or later get into an inconsistency by going through the following argument: "Suppose I ever believe that the native is a knight. Then I'll believe what he said—I'll believe that I don't (and never will) believe he's a knight. But also, if I ever believe he's a knight, then I'll believe that I *do* believe he's a knight (since I am normal), which means I would be inconsistent! Now, since I am consistent (sic!), then I can never believe that he's a knight. He said I never would. Hence he's a knight."

At this point the reasoner believes that the native is a knight. Being normal, he continues, "Now I believe he's a knight. He said I never would. Hence he is a knave."

At this point the reasoner is clearly inconsistent.

§5. Self-Fulfilling Beliefs and Löb's Theorem.

We now consider an advanced reasoner who is suffering from some ailment and visits a doctor whose word he trusts. He asks the doctor, "Will I ever be cured?" The doctor replies, "The cure is mainly psychological; the belief that you will be cured is self-fulfilling. If you ever *believe* that you will be cured, then you will be."

The reasoner leaves the doctor none too satisfied. Although he believes what the doctor has said, he cannot but ask himself, "But how do I know that I will ever *believe* that I will be cured?" He ponders this for awhile but gets nowhere, and so he decides to take a vacation on the Island of Knights and Knaves. One day he meets the Island Shaman. He knows that the shaman is either a knight or a knave, but he doesn't know which. Nevertheless, in a moment of desperation, he confides his case to the shaman and concludes by saying, "My doctor is trustworthy, and so if I ever believe that I'll get cured, then I will get cured. But I have no rational evidence that I'll ever believe that I'll get cured!" The shaman replies, "If you ever believe I'm a knight, then you will be cured."

The interesting thing is that it then follows that the reasoner *will* believe that he will be cured (and if his doctor was right, he will be).

Problem 4—After Löb's Theorem. How is this proved?

Solution. The solution will be more perspicuous if we give it partly in English and partly in symbolism. We let c be the proposition that the reasoner will be cured. The reasoner goes to the island already believing $Bc \supset c$. Then the shaman tells him $Bk \supset c$, where k is the proposition that the shaman is a knight, and so the reasoner believes $k \equiv (Bk \supset c)$.

The reasoner then reasons, "Suppose I ever believe that he's a knight. Then I'll believe what he said—I'll believe $Bk \supset c$. But if I believe he's a knight, I'll believe Bk (since I am normal). Once I believe Bk and $Bk \supset c$, I'll believe c. But if I believe c, then I really will be cured (as my doctor told me). Thus, if I ever believe that the shaman is a knight, then I'll get cured. Well, that's exactly what he said. Hence he's a knight!"

The reasoner, being normal, then continues, "Now I believe he's a knight, and I've already proved that if I ever believe he's a knight, I'll get cured. Hence I'll get cured."

At this point the reasoner believes that he will get cured (and if his doctor was right, then he will be).

I. Logicians Who Reason About Themselves

Exercise 1. Suppose in Problem 1 the native had instead said, "You *will* believe that I am a knave." Would the same conclusion follow?

Exercise 2. Same question with Problem 2.

Exercise 3. Suppose in Problem 3 the native had instead said, "You will believe I'm a knave before you believe I'm a knight." [The native asserts $B\sim k < Bk$]. Would the same conclusion follow?

Exercise 4. Suppose in Problem 4 the shaman had instead said, "If you ever believe that I'm a knight, then you will *believe* that you will get cured." Would that have helped the reasoner?

Exercise 5. Suppose, instead, the shaman had said, "You will believe that if I'm a knight, then you will be cured." Would that have helped the reasoner?

Exercise 6. Suppose the shaman had said, "You will never be cured and you *will* believe that I'm a knave." Would that have helped the reasoner? [Yes it would, why?]

Exercise 7. Suppose the shaman had said, "You will never believe I'm a knight and you will never believe that you will be cured." Would that have helped the reasoner?

Exercise 8. Suppose the reasoner had never consulted a doctor (and, hence, never had any prior belief that if he believes he will be cured, then he will be). Suppose the shaman had made two separate statements (1) "If you ever believe I'm a knight, then you will believe that you'll be cured" and (2) "If you ever believe I'm a knight, then you *will* be cured." Prove that the reasoner (of type 4) will believe that he will be cured.

Exercise 9. Here is a cute one. An advanced reasoner goes to the Island of Knights and Knaves and is told by a native, "You will believe that there is life on other planets." Prove that what the reasoner *will* believe is the following, "If he is a knight, then I'll believe that he is one."

Exercise 10. A native says to an advanced reasoner, "You will believe that if I am a knight, then you will believe that I am one." Prove that the reasoner will believe that the native is a knight.

II. *Incompleteness Arguments in a General Setting*

The preceding problems are obvious analogues of incompleteness theorems of earlier chapters. Actually, their solutions and the corresponding incompleteness theorems are special cases of the more abstract theorems to which we now turn.

We let $\mathcal{M}$ be a collection of the following items:

1. A set S whose elements we will call *sentences* or *propositions* (depending on the intended application) [They could also be elements of a Boolean Algebra.]
2. An element f of S called *falsehood*
3. A binary operation $\supset$ that assigns to every ordered pair $\langle X, Y \rangle$ of elements of S an element $X \supset Y$ of S
4. A subset of S whose elements will be called the *provable* elements of $\mathcal{M}$
5. A mapping B that assigns to every element X of S an element BX of S [Informally, BX means that X is a provable element.]

We shall call $\mathcal{M}$ an *abstract provability system*—or a *provability system*, for short. In application to the problems of Part I, the elements of S are propositions, and Bp is the proposition that the reasoner sooner or later believes p.

In application to the systems $\mathcal{S}$ of earlier chapters with a provability predicate $P(v_1)$, we associate with each such system $\mathcal{S}$ the following provability system $\mathcal{M}(\mathcal{S})$: The sentences of $\mathcal{M}(\mathcal{S})$ are those of $\mathcal{S}$, the provable sentences of $\mathcal{M}(\mathcal{S})$ are the provable sentences of $\mathcal{S}$, and for every sentence X we take BX to be the sentence $P(\overline{X})$.

Returning to abstract provability systems, we define a subset V of S to be a *valuation set* if $f \notin V$ and for any sentences X and Y, the sentence $X \supset Y$ is in V iff either $X \notin V$ or $Y \in V$. We call X a *tautology* if it belongs to every valuation set. A subset T of S will be called a *truth set* if it is a valuation set and if for every sentence X, the sentence BX is in T iff X is provable in $\mathcal{M}$. We define $\sim X$ to be $X \supset f$, and then the logical connectives $\wedge, \vee$ and $\equiv$ are defined in the usual manner.

We shall say that $\mathcal{M}$ is of *type 1* if the set of provable elements contains all tautologies and is closed under modus ponens (if X and $X \supset Y$ are both provable, then so is Y). We shall call $\mathcal{M}$ *normal* if for every provable X, the sentence BX is also provable. We shall call $\mathcal{M}$ *stable* if the converse holds (if BX is provable, then so is X). We shall call $\mathcal{M}$ *consistent* if f is not provable. We let consis be the

II. Incompleteness Arguments in a General Setting

sentence $\sim Bf$. A mapping Q from sentences to sentences will be called a *Rosser mapping* if for every sentence X, if X is provable, then so is QX, and if $\sim X$ is provable, then so is $\sim QX$.

We shall say that $\mathcal{M}$ is of *type 4* if for any sentences X and Y, the following conditions hold:

P_1: If X is provable, then so is BX ($\mathcal{M}$ is normal).
P_2: $B(X \supset Y) \supset (BX \supset BY)$ is provable in $\mathcal{M}$.
P_3: $BX \supset BBX$ is provable in $\mathcal{M}$.

Problems 1–5 of Part I, when stripped of their anthropomorphic setting, reduce simply to Theorem 1–5 below (which also generalize earlier theorems about first-order systems).

Theorem 1—After Tarski-Gödel. *Suppose there exists a truth-set T for $\mathcal{M}$ such that every provable element is in T, and suppose X is an element such that $X \equiv \sim BX$ is in T. Then neither X nor $\sim X$ is provable in $\mathcal{M}$ (yet $X \in T$).*

Theorem 2—After Gödel. *Suppose $\mathcal{M}$ is a normal system of type 1 and G is a sentence such that $G \equiv \sim BG$ is provable in $\mathcal{M}$. Then*

1. *If G is provable in $\mathcal{M}$, then $\mathcal{M}$ is inconsistent*
2. *If $\sim G$ is provable in $\mathcal{M}$, then $\mathcal{M}$ is either inconsistent or unstable.*

Theorem 3—After Rosser. *Suppose $\mathcal{M}$ is a system of type 1 and Q is a Rosser mapping for $\mathcal{M}$. Then for any sentence X, if $X \equiv \sim QX$ is provable in $\mathcal{M}$ and $\mathcal{M}$ is consistent, then neither X nor $\sim X$ is provable in $\mathcal{M}$.*

Theorem 4—After Gödel's Second Theorem. *Suppose $\mathcal{M}$ is of type 4 and there is a sentence G such that $G \equiv \sim BG$ is provable in $\mathcal{M}$. Then, if $\mathcal{M}$ is consistent, the sentence consis (i.e., the sentence $\sim Bf$) is not provable in $\mathcal{M}$.*

Theorem 5—After Löb. *Suppose $\mathcal{M}$ is of type 4, $BX \supset X$ is provable in $\mathcal{M}$, and there is a sentence Y such that $Y \equiv (BY \supset X)$ is provable in $\mathcal{M}$. Then X is provable in $\mathcal{M}$.*

These five theorems just about sum up the most important results of earlier chapters. A few remarks, though, may be in order.

We first turn to a corollary of Theorem 2. Let us consider a provability system $\mathcal{M}$ and, in addition, a mapping Φ that assigns to every sentence X and every natural number n a sentence $\Phi(X, n)$.

[Heuristically, we think of the provable sentences as being proved at various *stages*, and $\Phi(X, n)$ expresses the proposition that X is proved at stage n.] Let us say that Φ *enumerates* the set of provable sentences or that Φ is an *enumeration map* for $\mathcal{M}$ if for every sentence X, if X is provable, then $\Phi(X, n)$ is provable for at least one n. If X is not provable, then $\sim \Phi(X, n)$ is provable for *every* n. Let us call Φ an *adequate* enumeration map for $\mathcal{M}$ if, in addition, for every sentence X and every natural number n, if $\Phi(X, n)$ is provable, then so is BX. Let us say that $\mathcal{M}$ is ω-*consistent* with respect to Φ if for every sentence X, if BX is provable, then there is at least one n such that $\sim \Phi(X, n)$ is not provable. Then we have

Theorem 2A. *Suppose $\mathcal{M}$ is of type 1 and Φ is an adequate enumeration map for $\mathcal{M}$. Suppose G is a sentence such that*

$$G \equiv \sim BG$$

is provable in $\mathcal{M}$. Then

1. *If $\mathcal{M}$ is consistent, then G is not provable in $\mathcal{M}$.*
2. *If $\mathcal{M}$ is ω-consistent with respect to Φ, then $\sim G$ is not provable in $\mathcal{M}$.*

We remark that the adequacy of Φ implies that $\mathcal{M}$ is normal, and the ω-consistency of $\mathcal{M}$ with respect to Φ implies that $\mathcal{M}$ is stable. So Theorem 2A is a corollary of Theorem 2.

We note that the solution of Problem 2A is a special case of Theorem 2A, taking $\Phi(X, n)$ to be $B_n X$ (the reasoner believes X on the nth day).

Let us also note that the solution of Exercise 5 of Chapter VIII is a special case of Theorem 2A. We are given a formula $F(x, y)$ that enumerates in S the set of Gödel numbers of the provable sentences. We take $\Phi(X, n)$ to be the sentence $F(\overline{X}, \overline{n})$ and apply Theorem 2A.

Theorem 3 generalizes the solution of Exercise 6 of Chapter VIII. We take QX to be the sentence $\exists y(F(\overline{x}, y) \supset (\forall z \leq y) \sim G(\overline{x}, z))$. Then $\sim QX$ is logically equivalent to the sentence

$$\forall y(F(\overline{x}, y) \supset (\exists z \leq y) G(\overline{x}, z)).$$

As in Chapter IX, the proofs of Theorems 4 and 5 above are facilitated by first showing that from properties P_1, P_2 and P_3, it follows that for any sentences X, Y and Z, the following conditions hold:

P_4: If $X \supset Y$ is provable in $\mathcal{M}$, then so is $BX \supset BY$.
P_5: If $X \supset (Y \supset Z)$ is provable in $\mathcal{M}$, then so is $BX \supset (BY \supset BZ)$.

P_6: If $X \supset (BX \supset Y)$ is provable in $\mathcal{M}$, then so is $BX \supset BY$.

Properties P_1 and P_6 are the crucial ones for the proofs of Theorems 4 and 5 (cf. Exercises 11 and 12 below). Let us note that Kreisel's observation (page 103) still holds good in our present abstract setting. Theorem 4 is that special case of Theorem 5 in which $X = f$.

Exercise 11. Show that if $\mathcal{M}$ is of type 1 and has property P_6, then it has property P_3.

Exercise 12. Call $\mathcal{M}$ of type 4^- if it is of type 1 and has Properties P_1 and P_6 (it will then also have property P_3, by the above exercise, but may fail to have Property P_2). Show that Theorems 4 and 5 hold under the weaker assumption that $\mathcal{M}$ is of type 4^-.

III. Systems of Type G

We know by Löb's theorem for Peano Arithmetic that for any sentence X of P.A., if $P(\overline{X}) \supset X$ is provable in P.A., then so is X. This means that $P(\overline{P(\overline{X}) \supset X}) \supset P(\overline{X})$ is a *true* sentence of arithmetic. Well, it is not only true, but even provable in P.A., as we will see.

We define a provablity system $\mathcal{M}$ to be of *type G* if it is of type 4 and if for every sentence X, the sentence $B(BX \supset X) \supset BX$ is provable in $\mathcal{M}$.

We will say that $\mathcal{M}$ is *reflexive* if for every sentence X, there is a sentence Y such that $Y \equiv (BY \supset X)$ is provable in $\mathcal{M}$ [The system P.A. is reflexive since it is diagonalizable]. Let us say that $\mathcal{M}$ has the Löb property if for every sentence X, if $BX \supset X$ is provable in $\mathcal{M}$, then so is X. We now wish to prove

Theorem 6. *For any system $\mathcal{M}$ of type 4, the following three conditions are equivalent:*

C_1: $\mathcal{M}$ *is reflexive.*
C_2: $\mathcal{M}$ *has the Löb property.*
C_3: $\mathcal{M}$ *is of type G*

Of course, we already know that C_1 implies C_2 (by Theorem 5). To show that C_2 implies C_3, we first prove a lemma. For any sentence X, we let X^* be the sentence $B(BX \supset X) \supset BX$.

Lemma. *If $\mathcal{M}$ is of type 4, then $BX^* \supset X^*$ is provable in $\mathcal{M}$.*

Proof of Lemma. Suppose $\mathcal{M}$ is of type 4. For any sentence X we have to show that

$$B(B(BX \supset X) \supset BX) \supset (B(BX \supset X) \supset BX)$$

is provable in $\mathcal{M}$. It suffices to show that

$$(B(B(BX \supset X) \supset BX) \wedge B(BX \supset X)) \supset BX$$

is provable. Let

$$Y = B(B(BX \supset X) \supset BX) \wedge B(BX \supset X).$$

We are to show that $Y \supset BX$ is provable. Well, the following sentences are all provable.

(1) $Y \supset B(B(BX \supset X) \supset BX)$ (obvious)
(2) $Y \supset B(BX \supset X)$ (obvious)
(3) $B(BX \supset X) \supset BB(BX \supset X)$ (by P_3)
(4) $Y \supset BB(BX \supset X)$ (from (2) and (3))
(5) $Y \supset BBX$ (from (1), (4), using P_2)
(6) $B(BX \supset X) \supset (BBX \supset BX)$ (by P_2)
(7) $Y \supset (BBX \supset BX)$ (by (2) and (6) and propositional logic)

(8) $Y \supset BX$ (by 5), (7) and propositional logic)

Proof of Theorem 6. Suppose $\mathcal{M}$ is of type 4. We will show that $C_1 \supset C_2 \supset C_3 \supset C_1$.

(1) We already know that $C_1 \supset C_2$.

(2) Suppose C_2—i.e., that $\mathcal{M}$ has the Löb property. For any sentence X, the sentence $BX^* \supset X^*$ is provable in $\mathcal{M}$ (by the above lemma). Then, since $\mathcal{M}$ has the Löb property, X^* is provable in $\mathcal{M}$—i.e. $B(BX \supset X) \supset BX$ is provable in $\mathcal{M}$. Thus $\mathcal{M}$ is of type G.

(3) Suppose C_3—i.e. that $\mathcal{M}$ is of type G.

Now, $X \supset (BX \supset X)$ is a tautology; hence it is provable. Then by Property P_4, $BX \supset B(BX \supset X)$ is provable. Also

$$B(BX \supset X) \supset BX$$

is provable (by hypothesis) and so $BX \equiv B(BX \supset X)$ is provable. Then (by propositional logic)

$$(BX \supset X) \equiv (B(BX \supset X) \supset X)$$

is provable. Thus $Y \equiv (BY \supset X)$ is provable, where Y is the

III. Systems of Type G

sentence $BX \supset X$. Thus $\mathcal{M}$ is reflexive.

Remarks. 1. We passed from C_3 to C_2 via C_1. A simpler proof that C_3 implies C_2 is this: Suppose $\mathcal{M}$ is of type G. Suppose also $BX \supset X$ is provable in $\mathcal{M}$. Then so is $B(BX \supset X)$. But

$$B(BX \supset X) \supset BX$$

is provable, and so BX is provable. Since $BX \supset X$ is provable, so is X.

2. We passed from C_1 to C_3 via C_2. It might be instructive to look at a direct proof.

Suppose there is some Y such that $Y \equiv (BY \supset X)$ is provable. Then the following sentences are provable.

(1) $Y \supset (BY \supset X)$
(2) $(BY \supset X) \supset Y$
(3) $BY \supset BX$ (from (1), using Property P_6)
(4) $(BX \supset X) \supset (BY \supset X)$ (from (3) by propositional logic)
(5) $(BX \supset X) \supset Y$ (from (4) and (2))
(6) $B(BX \supset X) \supset BY$ (from (5), using property P_4)
(7) $B(BX \supset X) \supset BX$ (from (6) and (3))

[One can also prove the stronger fact that for $\mathcal{M}$ of type 4, for any sentences X and Y, the sentence

$$B(Y \equiv (BY \supset X)) \supset (B(BX \supset X) \supset BX)$$

is provable in $\mathcal{M}$[1]]

Exercise 13. Consider a provability system $\mathcal{M}$ and a mapping $\varphi(x,y)$ from ordered pairs of sentences to sentences. Let us say that $\varphi(x,y)$ has the *fixed-point property* (for $\mathcal{M}$, understood) if for every sentence Y, there is a sentence X such that the sentence $X \equiv \varphi(X,Y)$ is provable in $\mathcal{M}$. Now consider the following mappings, φ_1–φ_6:

$$\varphi_1(X,Y) = BX \supset Y$$
$$\varphi_2(X,Y) = BX \supset BY$$
$$\varphi_3(X,Y) = B(X \supset Y)$$
$$\varphi_4(X,Y) = B \sim X \wedge \sim Y$$
$$\varphi_5(X,Y) = B \sim X \wedge \sim BY$$
$$\varphi_6(X,Y) = \sim B(X \vee Y)$$

---

[1] cf. Boolos [1979] or Smullyan [1987].

To say that X is *reflexive* is to say that φ_1 has the fixed-point property. Now, suppose $\mathcal{M}$ is of type 4.

(a) Show that if $\mathcal{M}$ is reflexive, then every one of the mappings φ_1–φ_6 has the fixed point property.
(b) Show that if *any* one of the mappings φ_1–φ_6 has the fixed-point property, then $\mathcal{M}$ is reflexive (and hence of type G).

Exercise 14. How does (b) of Exercise 13 solve Exercises 4–7?

IV. Modal Systems

Let us now say just a little as to how all this is related to systems of *modal logic*.

Modal logic was originally developed for the purpose of explicating the notion of *necessary* truth (as opposed to merely contingent or factual truth), and the symbol $\Box$ was used to mean "it is necessary that". We are interested, however, in the more recent "provability" interpretation of modal logic in which $\Box$ is interpreted as "it is provable that".

The alphabet of modal logic (in the formalism that we shall follow) consists of a denumerable set of symbols called *propositional variables* together with the following five symbols:

$$\Box \supset \perp (\ )$$

The symbol $\perp$ (pronounced "eet"—it is a $\top$ written upside down) is thought of as denoting logical falsehood (it's synonymous with "f", as we have used it following Church [1956]).

The class of *modal formulas* is defined inductively by the following rules.

(1) $\perp$ is a modal formula, and so is each propositional variable.
(2) For any modal formulas X and Y, the expression $(X \supset Y)$ is a modal formula.
(3) For any modal formula X, the expression $\Box X$ is a modal formula.

One defines $\sim X$ as $X \supset \perp$, and the other logical connectives $\wedge, \vee$ and $\equiv$ are then defined as usual.

There are three modal axiom systems of increasing strength—K, K_4 and G—which are of particular interest for the study of prov-

IV. Modal Systems

ability predicates (or more generally, of abstract provability systems). The basic system K has as its axioms

A_1: All tautologies
A_2: All formulas of the form $\Box(X \supset Y) \supset (\Box X \supset \Box Y)$.

The system K_4 has the above axioms plus

A_3: All formulas of the form $\Box X \supset \Box\Box X$.

The system G has the above axioms plus

A_4: All formulas of the form $\Box(\Box X \supset X) \supset \Box X$.

The inference rules for these systems are *Modus Ponens* (from X and $X \supset Y$ to infer Y) and *Necessitation* (from X to infer $\Box X$).

We have associated with each first-order system S of arithmetic an abstract provability system $\mathcal{M}(S)$ and we can do the same with systems of modal logic. Given an arbitrary axiom system M whose formulas are those of modal logic, we define $\mathcal{M}(M)$ as that provability system whose sentences are the formulas of modal logic and whose provable sentences are the formulas provable in M, and whose mapping B is that which assigns to each modal formula X the formula $\Box X$. Then it is obvious that $\mathcal{M}(K_4)$ is a system of type 4 and $\mathcal{M}(G)$ is a system of type G. Any theorems about all systems of type 4 apply in particular to the modal system K_4—and, similarly, with systems of type G and the modal system G.

Going in the other direction, any theorem about K_4 gives us information about *all* systems $\mathcal{M}$ of type 4—and similarly with G and all systems $\mathcal{M}$ of type G. More explicitly, for any abstract provability system $\mathcal{M}$, we define a *translation* (of modal formulas) into $\mathcal{M}$ as a mapping φ that assigns to every modal formula X a sentence $\varphi(X)$ such that:

1. $\varphi(\bot) = f$;
2. $\varphi(X \supset Y) = \varphi(X) \supset \varphi(Y)$;
3. if $\varphi(X) = Y$ then $\varphi(\Box X) = BY$.

An obvious induction argument shows that any mapping φ_0 of all *propositional variables* to sentences of $\mathcal{M}$ can be extended to one and only one translation φ (of all modal formulas) into $\mathcal{M}$. By a *translate* of a modal formula X into $\mathcal{M}$, we mean any sentence $\varphi(X)$ where φ is any translation into $\mathcal{M}$. Then it is easy to see that if $\mathcal{M}$ is of type 4, then for any modal formula X, if X is provable in K_4, then all its translates are provable in $\mathcal{M}$. Similarly with the modal system G and systems $\mathcal{M}$ of type G. In particular, if X is provable in G, then all its translates into P.A. are provable in P.A. A remarkable result of Robert Solovay (known as his *completeness theorem for G*) states the converse—i.e., if all translates of X into P.A. are provable in P.A., then X is provable in G. Proofs of this can be found, e.g., in Boolos [1979], in which the whole theory of the modal system G (a big subject these days) is developed in great depth.[2]

In Smullyan [1987], we introduced the notion of *self-referential* interpretations of modal systems. By a modal *sentence*, we mean a modal formula in which no propositional variables appear. [These are called *letterless* sentences in Boolos [1979].] Thus all (letterless) sentences are built from the three symbols $\Box$, $\supset$ and $\bot$ and the two parentheses. We now define a modal sentence X to be *true* for a modal system M if it is true when we interpret $\Box$ to mean provability in M. More precisely, we define *true for M* by the inductive rules:

1. $\bot$ is not true for M;
2. $X \supset Y$ is true for M iff either X is not true for M or Y is true for M;
3. $\Box X$ is true for M iff X is provable in M.

We then call a modal system M *self-referentially correct* if every sentence provable in M is true for M. It is not difficult to prove that the systems K, K_4 and G are each self-referentially correct (a proof can be found in Smullyan [1987]). Since G is self-referentially correct, it immediately follows that G is consistent—and also that G is stable (since the provability in G of $\Box\Box X \supset \Box X$ implies its truth for G, which means that if $\Box X$ is provable in G, then so is X. And the same argument goes for K_4). Under the self-referential interpretation of G, the sentence $\sim \Box \bot$ expresses the *consistency* of G (it is true for G iff G is consistent). Since G is consistent, the sentence $\sim \Box \bot$ is, indeed, true for G but not provable in G (since G itself is a system of type G). This system G is, thus, a very simple example of a consistent system that cannot prove its own consistency.

Exercise 15. Prove that there is no formula X such that $\sim \Box X$ is provable in G.

---

[2] Before turning to this volume, the reader might first consult Chapter 27 of Boolos and Jeffrey [1980]. Also, those readers who are interested in the project of regarding "belief" as a modality (as we did in Part I of this chapter) might take a look at our volume [1987], or [1986], in which an informal presentation is combined with a formal one.

IV. Modal Systems

Exercise 16. Let us say that a provability system $\mathcal{M}$ can prove its own stability if for every sentence X of $\mathcal{M}$, the sentence $BBX \supset BX$ is provable in $\mathcal{M}$.

Show that if $\mathcal{M}$ is a consistent stable system of type G, then $\mathcal{M}$ cannot prove its own stability. [In particular, not all formulas of the form $\Box\Box X \supset \Box X$ are provable in the modal system G.]

References

[1] Askanas, Malgorzata. *Formalization of a semantic proof of Gödel's incompleteness theorem.* (Doctoral dissertation), Graduate Faculty in Mathematics, The City University of New York. (Abstract J. of Symbolic Logic, vol. 42, p. 154), 1975.

[2] Boolos, George. **The Unprovability of Consistency.** Cambridge University Press, 1979.

[3] Boolos, George and Richard Jeffrey. **Computability and Logic**, (second edition). Cambridge University Press, 1980.

[4] Carnap, Rudolph. **Logische Syntax der Sprache.** Wien (1934), p. 91.

[5] Church, Alonzo. **Introduction to Mathematical Logic**, vol. 1. Princeton University Press, 1956.

[6] Ehrenfeucht, A. and S. Feferman. *Representability of recursively enumerable sets in formal theories.* **Archiv für Mathematische Logik und Grundlogenforshung**, vol. 5 (1960), p. 37–41.

[7] Gödel, Kurt. *Uber formal unentscheidbare Sätze der Principia Mathematica und Verwandter, Systeme I.* **Montshefte für Mathematik und Physics**, vol. 38 (1931), p. 173–198.

[8] Henkin, Leon. *A problem concerning provability.* **J. Symbolic Logic**, vol. 17 (1952), p. 160.

[9] Hilbert, D. and P. Bernays. **Grundlagen der Mathematik.** Berlin, Springer, vol.1 (1934), vol.2 (1939).

[10] Kalish, D. and R. Montague. *On Tarski's Formalization of predicate logic with identity.* **Arch. f. Math. Logik und Grundl.** vol. 7 (1964), p. 81–101.

[11] Kleene, Stephen Cole. **Introduction to Metamathematics.** D. Van Nostrand Company, Inc. 1952.

[12] Löb, Martin H. *Solution to a problem of Leon Henkin.* **J. of Symbolic Logic** (1955), p. 115–118.

[13] Mostowski, Andrzej. **Sentences Undecidable in Formalized Arithmetic.** North Holland Publishing Company. 1952.

[14] Myhill, John. *Creative sets.* **Zeitschrift für mathematischer**

References

Logik und Grundlagen der Mathematik, vol. 1 (1955), p. 97–108.

[15] Putnam, Hilary and Raymond Smullyan. *Exact separation of recursively enumerable sets within theories.* **Proceedings of the American Mathematical Society**, vol. 11, p. 574–577.

[16] Quine, Willard Van Orman. **Mathematical Logic.** Norton, 1940.

[17] *Concatenation as a basis for arithmetic.* **J. Symbolic Logic**, vol. 11 (1946), p. 105–114.

[18] Robinson, Raphael M. *An essentially undecidable axiom system.* **Proceedings of the International Congress of Mathematicians**, vol. 1 (1950), p. 729–730.

[19] Rosser, J. Barkley. *Extensions of some theorems of Gödel and Church.* **J. of Symbolic Logic**, vol. 1 (1936), p. 87–91.

[20] Shepherdson, John. *Representability of recursively enumerable sets in formal theories.* **Archiv für Mathematische Logik und Grundlagenforshung**, (1961), p. 119–127.

[21] Schoenfield, J.R. *Undecidable and creative theories.* **Fundamenta Mathematica**, vol. XLIX (1961), p. 171–179.

[22] Schoenfield, J.R. **Mathematical Logic.** Addison Wesley, 1967.

[23] Smullyan, Raymond M. *Theory of formal systems.* **Annals of Mathematics Studies**, (1961), p. 47.

[24] *Logicians who reason about themselves.* **Reasoning About Knowledge, Proceedings of the Conference.** Morgan Kaufmann Publishers, (1986), p. 341–352.

[25] Smullyan, Raymond M. **Forever undecided: A Puzzle Guide to Gödel.** Alfred A. Knopf, 1987.

[26] Smullyan, Raymond M. **Recursion Theory for Metamathematics.** Oxford University Press, 1992.

Index

A^*, 6, 26
Abstract Forms of Gödel's and Tarski's Theorems, 5
Arithmetic (note the capital "A"), 19
arithmetic (note the small "a"), 19
Askanas' Theorem, 114
axiomatizable, 57

Carnap's rule, 113
complete representability, 97
concatenation to the base b, 20
consis, 108
constructive arithmetic relations, 41
correctly decidable, 66

degree, 16
designation, 17
diagonal function, 26, 102
diagonalizable, 108
diagonalization, 6, 24, 26
$d(x)$, 6, 26, 102

eet, 132
E_n, 23
$E[\overline{n}]$, 25
$(\exists v_i \leq c)F$, 41
$(\exists v_i \leq t)F$, 17
express, 19
expressibility, 5
expressible, 6
extension, 57

fixed point, 102
$(\forall v_i \leq t)F$, 17
$(\forall x \in y)$, 33
formation sequence, 33

Gödel Numbering, 22
Gödel Sentences, 8, 24
Gödel-Rosser Incompleteness Theorem, 76
Gödel's Incompleteness Theorem for P.E., 36

iff, 6

K_{11}, 32
Kripke, Saul, 46

$\ell_b(n)$, 21
$\ell(n)$, 21
language $\mathcal{L}_E$, 14
$\leftrightarrow$, 6

Myhill, John, 43

$\mathcal{N}$, 57
nameable, 6
neg(x), 34

ω-rule, 113
ω-consistent, 57
ω-incompleteness, 73
ω-inconsistent, 56

P^*, 59
P.A., 40
P.E., 28

Index

P_E, 36
Peano Arithmetic, 40
Peano Arithmetic with Exponentiation, 28
$\Pi(x, y, z)$, 87
provability predicates, 106

(Q), 68
(Q_0), 69

R^*, 59
(R), 69
(R_0), 69
R_E, 36
r.e. systems, 57
recursive relations, 98
recursive set, 98
recursively axiomatizable, 57
regular formula, 16
represent, 58
representation function, 26
Robinson, Raphael, 69
Rosser system, 78
Rosser system for n-ary relations, 78
Rosser system for sets, 78
Rosser's Undecidable Sentence, 81
$r(x, y)$, 25

$Sb(x)$, 34
separability, 77
separates, 77
Separation Lemma, 79
sequence number, 32
$Seq(x)$, 32
Shepherdson's Representation Lemma, 88
Shepherdson's Representation Theorem, 86
Shepherdson's Separation Lemma, 91
Σ_0-complete, 66

Σ_0-formulas, 41
Σ_0-relations, 41
Σ_1-formula, 42
Σ_1-formulas and relations, 42
Σ_1-Relations, 42
Σ-formulas, 42
simply consistent, 56
Strong Definability of Functions in $\mathcal{S}$, 98
strong separation, 87
strongly separates, 87
subsystem, 57
superset, 12
$s(x)$, 34
system (Q), 69
system (R), 69
system (Q_0), 69

Tarski's rule, 113
Tarski's Theorem, 24, 27
Truth in $\mathcal{L}_E$, 17
truth predicates, 104

undecidable sentences, 10

$Var(x)$, 34

weak separation, 87
weakly separates, 87

$x \in y$, 32
$x \underset{z}{\prec} y$, 32
$x \exp y$, 34
$x \text{ id } y$, 34
$x \text{ imp } y$, 34
$x \text{ le } y$, 34
$x \text{ pl } y$, 34
$x \text{ tim } y$, 34

Printed in the United Kingdom
by Lightning Source UK Ltd.
103078UKS00001BB/90